PARADISE
FOUND

PARADISE FOUND

A HIGH SCHOOL FOOTBALL TEAM'S RISE FROM THE ASHES

BILL PLASCHKE

WM

WILLIAM MORROW

An Imprint of HarperCollins*Publishers*

HarperCollins books may be purchased for educational, business, or sales promotional use. For information, please email the Special Markets Department at SPsales@harpercollins.com.

FIRST EDITION

Designed by Nancy Singer

Library of Congress Cataloging-in-Publication Data has been applied for.

ISBN 978-0-06-301451-0

21 22 23 24 25 LSC 10 9 8 7 6 5 4 3 2 1

For the Marvelous Mary Plaschke

Thanks, Mom

CONTENTS

PARADISE
FOUND

PROLOGUE

The e-mail hit my in-box on January 31, 2019. It was from Gary Pine, athletic director at Azusa Pacific, a small Christian university located just outside Los Angeles.

> Bill . . . I have a story idea for you to chew on for a while and see if you want to follow through with sometime this winter/spring.
>
> Rick Prinz is the head football coach at Paradise High, which, as you know, was the city that was destroyed by the Camp Fire back in November. He has a connection to some people at Azusa Pacific, and we've asked him to come speak to about 1,000 high school students in March to tell his story.
>
> In preparation for his March speaking, I spoke with him yesterday and realized he may have a story for you, kind of a "rest of the story" concerning his football team, the devastation, and his own personal story.

As a sports columnist for the *Los Angeles Times* who has long loved telling these sorts of stories, I was immediately entranced by this one and quickly e-mailed Rick Prinz with a request to chronicle the comeback of his team. I didn't hear from him for three weeks.

"Hey, Bill, sorry I'm just getting back to you," Prinz said when he finally called. "The last couple of months, I've been busy just trying to live."

PARADISE WAS, FOR MANY, PARADISE. The tiny Northern California town, located eighty-seven miles north of Sacramento and fifteen minutes northeast from Chico, is nestled amid the tall pines and sweeping vistas of the Sierra Nevada foothills. It was mostly a bedroom and retirement community; a quiet, picturesque place to raise a family or spend one's golden years. The median age was fifty, with folks scattered in both small cabins and expansive ranch homes tucked away on winding, wooded streets.

The main drag, Skyway Road, was lined with an eclectic collection of small businesses and shops. There was a hospital, and there was a Starbucks, but folks would drive down to Chico for the serious shopping, and they liked it that way. Paradise was a wondrous daily escape. It was a sharp and steep turn off the beaten path, and the townspeople loved and cherished its isolation and its privacy. Their only nagging fear, because of the many trees and high winds, was always fire. Over the years, there had been many small fires, many evacuations, and many worries that one day they would be burned by a big one.

That day was the morning of November 8, 2018. In the deadliest and most destructive blaze in California history, the Camp Fire roared through Paradise, killing 86 people and virtually leveling the city.

Paradise's population before the fire was 26,800. Its population after the fire was approximately 2,034.

More than 18,800 structures were destroyed, including nearly 14,000 homes. Roughly 30,000 people were left homeless.

The fire was caused by an aging and faulty electrical transmission line owned and operated by Pacific Gas and Electric Company. The utility would reach a $13.5 billion settlement with residents and later plead guilty to eighty-four counts of involuntary manslaughter.

The above paragraphs do not come close to addressing the true horror of the situation. The flames took all of four hours to engulf the town. Residents literally ran for their lives. Their cars jammed Skyway Road, the main artery out of town, causing a flaming traffic snarl in which some were burned alive in their vehicles. On this day, it took more than five hours to drive down to Chico, as honking cars full of screaming people rolled slowly between two walls of fire under a midmorning sky that was as black as midnight. Prinz was one of those people. Most of his teenage football players were, too.

When the smoke cleared eventually, one of the few remaining major structures in town was Paradise High School. It is a wooded campus of several main buildings that slopes down to a number of sports fields. Somehow it mostly escaped the flames. And somehow, sitting virtually untouched two levels below the classrooms, was the school's crown jewel: a sixty-year-old football stadium named after a former legendary Paradise High coach from the 1970s and 1980s, the unflinching Om Wraith Field.

Was that a sign? The current coach didn't think so. Two months after the fire, sitting at a cluttered desk in a cavernous concrete wing of the school's temporary home in a warehouse at Chico Airport, Rick Prinz calculated loss. Of the 104 players in his program, 95 had lost their homes. All of his eight coaches lost their homes. His varsity squad had been whittled from 76 players to 22 because the majority of families had either left the area or enrolled in schools closer to their temporary homes. The boys' equipment had melted, their uniforms burned. They didn't even have a football.

When we finally connected over the phone in the spring of 2019, Prinz told me, "I don't know how we're going to have a team, I don't know how we're going to play a schedule, I don't even know if I'm going to have a job."

But he knew what he had to do. From that lone remaining field, he would rebuild. With his devastated team, he would try to resurrect the spirit of his destroyed town.

"We don't have a choice," he said intently. "The kids need this. The town needs this. Everyone around here needs hope. We're going to try to give them football. We're going to try to be that hope."

So, in the face of unrelenting obstacles and against all common sense, they tried to play football.

This is their story.

ONE

The Truck

NOVEMBER 8, 2018

*Long is the way, and hard, that out
of Hell leads up to light.*

—John Milton, *Paradise Lost*

The plan is to practice at 3:00 today. If it is to Smokey, we will modify our activity. I will keep you informed if anything changes.

The text was sent on Thursday, November 8, 2018, at 8:10 a.m. from coach Rick Prinz to the seventy-six members of the Paradise High Bobcats varsity football team. It wasn't his finest literary moment. There were misspellings. There were typographical errors. He hoped his team would forgive him. As he typed, literally all hell was breaking loose.

Prinz had followed his usual routine that morning, the same sort of routine that had marked the sixty-year-old former high school middle linebacker's twenty years as the leader of the football team representing Paradise. His simple, scheduled life suited him well in this mountain village of about twenty-five thousand people. The locals would bring him their young men, he would teach them all the same and treat them all the same, while relying on the same playbook for twenty years, the same fighting drills, the same climb-on-the-goalpost-for-thirty-seconds punishment if you took off your helmet, and the same hugs when he would send them off into the world—whether to community college or fire school or apprenticeship as an electrician.

And the same style of winning, too. In the last two decades, Coach Prinz's Paradise Bobcats had claimed ten league titles, six sectional championships, and an overall record of 166-54.

On that chilly November morning, Prinz—stout, balding, white goateed; the portrait of stability—figured he knew exactly what he was doing because he had done it for so long.

He awoke in his cedar home at five thirty in the morning, drove his Toyota Tacoma truck ten minutes down a winding, narrow road to the cluttered campus of about a thousand students, parked in the back next to the gym, stepped into the small fitness room, and, by six, he was working the elliptical machine while reading the Bible on his smartphone. With his steely eyes and soft smile, the former youth pastor often coached through inspirational sayings, and it was from one of those Bible mornings that he was inspired to type a pep talk to his team three days before. The previous Friday might, they had finished the regular 2018 season with a 38–6 trouncing at the hands of rival Pleasant Valley High School, and so even though they were 8-2 and entering the playoffs against the relatively weak Red Bluff Spartans, they looked tired and beaten and needed a jolt. Rick gave that to them with an eloquent note on the Hudl sports sharing app.

> Word of the Week: Resilience—the capacity to recover quickly from difficulties, toughness. The inability to quit. Are you resilient? Are you going to step up? How will you prove it?

That was on Monday. This was Thursday. As he went from pumping his legs on the elliptical to pumping iron, then showered in his cramped coach's office he shared with physical education teacher Seth Roberts, Prinz had no idea that his sermon would soon take on new meaning.

At seven fifteen, Prinz and Roberts drove to the local Starbucks, several blocks away. He ordered his usual dark roast with half-and-half. The sun was just rising over the ridge. The sky was painted in shades of orange and black. He noticed nothing out of the ordinary. However, while pulling past the drive-through window, he heard someone inside say that there was a fire in the canyon. Everyone in the coffeehouse was peering through glass windows in that direction.

He craned his neck to the sky and saw smoke. He shrugged. Small fires and bursts of smoke often rose from the rugged forests around Paradise. Besides, it appeared to be at least a dozen miles away. This was nothing. This was just another day.

When Prinz returned to the school at seven forty-five, he climbed out of his truck and looked into the sky. This time the plume of smoke was much larger. Some students and teachers were chattering about it. There was an uneasy buzz in the air. Prinz shrugged again

"Is that smoke, or are those just clouds?" he said to no one in particular. "That could be nothing."

He walked back into his office. The Red Bluff game, to be held at an undetermined neutral site, was barely twenty-four hours away. He needed to check his game plan. He needed to figure out how to motivate his kids to win the three playoff games required to capture another sectional championship.

This would not be any ordinary postseason. This would be a retirement gift. Unbeknownst to anyone but his wife, Veronica, Prinz had decided this was his final season, making for an even twenty. The demands of both teaching a full day of physical education and then working until late at night with the football team were finally too much. The majority of his coaching staff worked other jobs off campus, meaning he was solely responsible for not only what happened on the football field but also what happened with his players in the classroom. He was weary of dealing with grade eligibility issues, and player-teacher conflicts, and all the paperwork involved in fielding a team. He was tired, his team was coming off a career-worst 2-8 season in 2017, this 2018 rebirth would be his final one, he was done, and these playoffs would be his farewell tour.

So he was thinking only of Red Bluff at seven forty-five that morning when his office mate Roberts bolted out of his chair and ran toward the door. Seth's wife had called and ordered him home. Then another PE teacher ran past his office on her way to the parking lot. Then another. They asked him to handle their classes. Rick shrugged and agreed. He still didn't understand the reason for all the commotion.

Finally, he followed the other teachers outside to the front of the school, and it hit him. Literally: a glowing ember falling at his shoes. Then another one. And another one. He looked up, and the sky that had seemed so harmless had been replaced with roiling, black smoke that was slowly turning day to night. He was standing next to principal Loren Lighthall. For the school boss, the embers clinched it. For the football boss, it was business as usual.

"We're canceling school," said Lighthall.

"I'm not canceling practice," insisted Prinz.

The coach retreated to his office located inside the boys' locker room. He locked the office door. He locked the locker room. He locked the gym. He stepped outside to monitor all the students who were scrambling into cars or waiting for rides.

This was when he sent the shaky text that there would indeed be practice. Nineteen years of this. Red Bluff on Friday. The routine must not be broken. They had practiced in smoke. They had practiced in the snow. They had practiced in searing heat and torrential rain and all sorts of weather that floated through their mountain village. It was only eight o'clock. The smoke had time to clear. They were practicing today.

Prinz pushed the Send button on his phone and walked back to the front of the school. The smoke was growing thicker and wider. He thought about his five-month-old grandson, Dylan, who had recently been discharged from a Sacramento hospital after a three-month stay following his premature birth. His twenty-four-year-old son Seth, Seth's girlfriend Caro, and their baby Dylan had moved into his house. What if the smoke impaired Dylan's breathing? Maybe the baby should be moved to a safer place.

Prinz called Seth and suggested he think about evacuating. He told him not to panic—it wasn't that serious—but for Dylan's sake, maybe they should consider leaving for the day. Seth agreed. Prinz ended the call to finish monitoring the departing children amid the accumulating smoke.

Suddenly Principal Lighthall approached him in a panic. Suddenly it got real.

"The fire is at the Ponderosa School," says Lighthall.

Ponderosa? The elementary school just a few miles away?

In that moment, for the first time in forever, Rick Prinz abandoned his routine, threw away his wing-T playbook, and called a Hail Mary audible. For the first time, this immovable, unsinkable veteran football coach dropped everything and ran.

He left Lighthall in midsentence and sprinted toward his truck while pulling out his phone and calling Seth back.

"Pack up! Pack up!" shouted Prinz.

"I'm already packed!" shouted Seth.

"I'm coming home to help!"

Prinz climbed into his truck and drove uphill toward his house. Now he could see it, the end of morning, the beginning of midnight. Crackling embers falling everywhere. Then he turned onto Skyway Road and encountered another nightmare.

Traffic traveling downhill was bumper to bumper to bumper, with cars just lined up and stopped. An evacuation apocalypse. The road was clear going uphill toward his house, but not for long. In the distance, just beyond Coldren Road, the side street that led to his home, panicked drivers were beginning to occupy all lanes going downhill. Prinz had about a minute to make a left turn, or he would have effectively been blocked from his street by the oncoming traffic.

He floored it, squealing into the turn, and successfully pulled onto Coldren just as the approaching cars barely stopped short of hitting him. He breathed a sigh of relief, just long enough to pull into another hell. His quaint, dark-cedar home was surrounded by fire, trees burning, smoke billowing, and—suddenly—an explosion, then another, then another.

Seth and Caro, cradling their infant son, were getting into their 1995 truck. But before they could pull away, Prinz stopped them. What if that old truck wasn't strong enough to get them down the hill? Prinz's truck was newer: 2012.

"Move everything to my truck, you take my truck, and go, now!" Prinz instructed.

But there wasn't enough room for everyone in Rick's vehicle.

"What are you going to drive?" Seth shouted.

It was at this point that steadfast Rick Prinz once again decided to change the script. Also sitting in the driveway was a 1947 candy-apple-colored Ford truck that had belonged to his father, Bill. It was his father's reward to himself for returning home safely from the Korean War after escaping from a prisoner-of-war camp. The truck was Bill Prinz's constant companion during his many years working for the US Forest Service. Before suffering a fatal stoke in 2000 at age seventy, he had already willed the truck to Rick—with one request.

"Please don't ever let it go," he told his son. "I don't care if you need the money; I don't care if it's the last thing you own. Please keep my truck."

"I promise," said Prinz.

And for the last eighteen years, Prinz drove the truck once or twice a week around the neighborhood, keeping it shining, presciently even replacing the radiator and the fan the previous summer in case he actually ever needed to drive it out of Paradise. It was a promise kept. And now, on that black Thursday morning in November, looking for one shred of stability as this new world was quickly crumbling around him, it was promise remembered.

"I'm taking my dad's truck!" Prinz shouted to Seth as Seth drove away, with Prinz praying it would not be the last words he'd ever speak to his son.

Prinz ran inside the overheated house, grabbed a few boxes containing some photos of his four children, threw them in the back of the ancient truck, and scrambled inside. He drove to the end of Coldren Road, made a right turn into the solid wall of traffic, then noticed something just as frightening as the flames that were surrounding him.

The old truck's gas gauge was hovering at E, and it was good for only about a dozen miles to the gallon. It was not going to make it the twenty miles to safety, especially with the Skyway becoming a parking lot. There was no access to a gas station; they were either closed or crowded or burning. He would have to literally push the car as far as he could, hopefully to the top of Skyway Road, and then cruise down on fumes.

"This is the end of times," he told himself as he turned off the ignition, hopped out of the cab, and, with one hand on the drivers' side door and the other on the steering wheel, began to push.

By then, he had talked to everyone else in his family, and all were accounted for. His wife, Veronica, was at her job at a periodontist office in Chico, seemingly out of the reach of the fire. Oldest son Billy had already evacuated his apartment in lower Paradise. And his other two sons, Jacob and Taylor, lived in Chico. His widowed mother, Carole, lived in upper Magalia, north of Paradise in a neighborhood that would be untouched by the fires. Everyone seemed to be safe except for him and Seth, who called just as Prinz began putting his back into pushing the seventy-year-old vehicle.

"Dad, I'm a few miles down from you. Traffic isn't moving; fire is everywhere. What do I do?"

On Rick Prinz's worst day, this was his worst moment. His son needed him, and he couldn't be there to help. It was crunch time, and he was out of plays. The old football coach began to cry, softly.

"I don't know what to tell you to do, Son," he said. "I just don't know."

At that moment, several cars ahead, he saw a family in a truck make a right turn onto a bike path that wound down through town. The truck bounced off trees and over ridges as it disappeared into the smoldering woods. Another car followed, and another, and as Prinz pushed his truck into their spots, he wondered, "Should I follow them? Will this old truck survive those trees?" He could then see the car's taillights disappearing into flickering flames. There was no way he could follow them. They were driving into death. He couldn't go, but he couldn't stay. The explosions were getting louder, the flames were getting closer.

So, walking alongside the truck's cab, he continued to push and steer, push and steer, inches at a time, downhill through hell. The smoke-filled sky was black. Propane tank explosions from nearby homes and businesses filled the air. He kept pushing, but he was starting to panic. He was behind a recreational vehicle with a gas can on top of it. For a moment, Rick seriously considered climbing up there and stealing the gas. But no, no, he could never do that. If he was going to die, he was going to die honestly.

Thirty minutes passed. He moved two hundred feet. Thirty more minutes. He moved four hundred feet. His phone rang. It was Seth again. His son had found a side road that had since been engulfed in flames. He'd made it halfway down the hill. This was good news. Yet there was still more bad news.

"Dad, the Skyway down here is on fire," he said. "You're not going to make it."

"Son, I'll make it!" Prinz shouted before the cell service went dead.

He was surprised at the strength of his voice, when, in fact, he barely believed it himself. Seth certainly didn't believe it. Later that morning, when he finally made it to Chico and met his mother in the Raley's grocery store parking lot, he gave her the sad news.

"I don't think Dad's going to make it," Seth said, his voice trembling, and together they huddled and prayed for the coach on the hill.

BACK UP ON SKYWAY, PRINZ kept pushing, taking two and a half hours to advance just two miles. Was he pushing his father's truck through the fire or into the fire? He wasn't sure which, and he was too frightened to guess. Rick actually thought about texting his players that he was canceling practice—worried that some of them would drive back to the school to prepare for the Red Bluff game. That was how he coached them, right? But his phone still had no service. He was distraught that he had not called off practice earlier; he prayed nobody stuck around until it was too late. He prayed there would be no death on his hands.

Prinz wondered if it was too late for him. People behind him were leaving their vehicles in the middle of the road and running down the middle of the highway. One man shook his head at Prinz as he jogged past.

"Forget the truck, save your life!" he shouted.

At that moment, the truck, his father, the memories, that was the entirety of Prinz's life, and he wasn't going to give it up. Three times he stopped, closed the door, and prepared to abandon the heirloom. Three times he stopped himself.

"My dad escaped from a death march; I can't do this one thing for him?" he said to himself. "I'm not leaving this truck. And I'm not dying here."

He had an additional problem that for most would seem relatively insignificant but was huge for a coach with such visibility and pride. For the last two hours, Prinz had to urinate, but he didn't know where to go. He couldn't leave the truck, and he couldn't just go in the street with seemingly half the town surrounding him. Finally reaching a bursting point, he grabbed an empty Gatorade bottle from the truck and relieved himself into it. He thought, "The world really is coming to an end."

After three hours of manually moving his car, Prinz came to a point where the road flattened in the middle of Paradise and began to curve uphill. He knew he could no longer push. He had to drive on whatever fumes of gas remained. An extra lane had been opened through town, but the traffic was still inching at only about five miles per hour, and the truck was sputtering as he clutched the wheel and cringed. The fire was visible through the smoky darkness, shooting flames on both sides of the road. He saw houses burning. He saw businesses collapsing. He looked down one road and saw the home of junior running back Lukas Hartley burning to the ground. He passed the Paradise city sign, now just a smoldering heap. The old truck's cab had no air-conditioning, so it felt like it was a hundred degrees inside. He could barely breathe.

The road widened again, and, at last, traffic began moving, and Rick Prinz's mind began racing. Would the gas last? Or, after all this, would he have to abandon it to the fire? What would his father think? What would his father do? And, really, as he glared into the flames filling his rearview mirror, wasn't the horror just beginning? Were any of his players marooned in that fire? Would any of the town be left standing? His home was probably gone, his school was probably gone, his job was probably gone, and his football team was almost certainly going to be disbanded.

THESE NIGHTMARES PLAYED OVER AND over in his head until, finally, six hours after he'd first felt the falling embers, he rolled into the outskirts of Chico. But once at the bottom of the hill, the fire safely behind him, Prinz found himself in another horror show. The line of slowing cars was surrounded by people wandering up and down the lanes, tapping on windows and peeking into backseats. They were folks whose families had been stuck on the hill. They had lost contact when cell service died. They were looking for their loved ones, even in strangers' cars, crying and pleading and praying.

"Mom, are you in there?"

"Grandpa, is that you?"

Prinz became choked up again as he called his wife, who was waiting at her sister Mary's house in Chico.

"Seth said you weren't going to make it," said Veronica, her voice breaking.

"Well, I guess I made it," said Prinz, who then drove through the crowded streets to the home of in-laws Scott and Mary Jenkins.

Veronica was waiting on the porch. They hugged and cried, and Prinz was dragged into the house in exhaustion. Once he caught his breath, he thought again, What about his team? What about his players who may have stayed on the mountain too long because they thought there would be practice? His players trusted him, and he had let them down, and now he was fearful for their lives.

Prinz finally had wireless service, so he logged on to the team's message board and saw a string of notes beginning at around noon, which is when the first survivors started arriving at the bottom of the mountain.

"Practice canceled," wrote Trevor Curtis, a senior center.

"Everyone safe?" wrote senior running back Dominic Wiggins.

The answers over the next four hours varied from comforting to distressing. Many players wrote, "I'm good" or "I'm safe," while others offered gruesome details.

"Watched my truck blow up, and the house I just moved into, homeless again," wrote senior safety Austyn Swarts at 12:54 p.m.

"I'm good, making my way to Chico," wrote junior running back and defensive back Ethan Hughes at 12:56 p.m. "Pretty sure we lost our house."

"Pretty sure my game gear burned up," wrote junior safety Dylan Blood at 4:16 p.m.

"I'm sorry for everyone whose house just burnt down," wrote senior receiver Greggor McGrath at 4:18 p.m. "Mine did too, and now I have nothing."

Then there was the 5:42 p.m. message from senior lineman Clayden Blackwell, written in an unnerving stream of consciousness: "Was stuck in the middle of the fire at kmart for 7 hours. just got out like 40 mins ago. ran out of gas on the way you and got stuck but out safe. thank god give all the glory to him for me being here still."

Prinz read the messages with both sadness and relief. Like all of Paradise, his team had suffered a terrible trauma. But between him and his coaches, he could at least take heart knowing that all of his players survived. And, as the messages continued to flow, he noticed a change. The expressions of shock and despair were giving way to declarations of resilience. His kids were now writing about the Red Bluff playoff game, scheduled for the next day. They had no homes, no town, no clue about the future, but they still wanted to play.

"I want to play this game still our town needs it, football brings our town together," wrote senior quarterback Colby Cline at 4:14 p.m.

"Fuck ya it does that game needs to happen," agreed Trevor Curtis a minute later.

"I mean, I do too, the question is . . . how?" responded Ben Dees.

"Yeah, exactly," chimed in running back Trevor Thurman.

One minute later, Trevor Curtis weighed in with, "We'll find a way, boys."

Prinz read the messages and cringed. In their energetic innocence, his players had no idea of the obstacles that they would be facing for months, much less the next few days. He agreed with Ben Dees. He wanted to play, the kids deserved it, they'd had a great season, they could win a championship. The question was . . . how?

Prinz gulped hard and sent his team his first note since coming down the mountain.

"Finally figured out how to use Hudl message on my phone. Go figure. Anyway, great to hear from you guys," he wrote at 6:11 p.m. "We will figure football out. Right now the important thing is you are safe. Sorry about those who have lost homes. I'm not sure about mine. Stay strong! Love you guys! Proud of each of you!"

He put down his phone, staggered back to an air mattress on the floor of his in-laws' home, hugged Veronica again, and tried, in vain, to fall asleep. He was still tossing around on Friday morning when he received news from Anne Stearns, Paradise High's young, new athletic director. She had arranged options. The game could be postponed until Saturday and played at Red Bluff High School. If that wasn't going to work, the Spartans generously agreed to forfeit the game and allow Prinz's Bobcats a week to prepare for their second-round opponent. He had two choices, both involving saving the game and inspiring the players and perhaps soothing the town. But then he talked to his coaching staff and wondered if, realistically, he had any choice at all.

Andy Hopper, the offensive line coach and the team's inspirational leader, had lost his house and all his possessions, with no insurance. Defensive coordinator Paul Orlando lost not only his home, but also five family members had seen their homes go up in flames, too. Defensive assistant coach Nino Pinocchio had lost everything, as had kicking coach Jeff Marcus, receivers coach Shannon Magpusao, and defensive line coach Bobby Richards. And Prinz himself didn't know yet whether he had lost everything. The rubble-filled town was shut down, and he couldn't get any information about whether or not his house had survived.

"How can I make a decision about a football game?" he asked Veronica. "I have coaches on couches, players on floors, nobody knows where they are going to live or how they are going to live. Yet everyone still wants to play. What do I do?"

All Friday morning, Prinz kept putting off the decision, hoping for some divine intervention. He was just a football coach working on a $4,000-a-year stipend. Why was he in charge of making such a big decision that would affect so many people? Couldn't someone else make the call?

Prinz couldn't give in. He never gave in. He couldn't tell his seniors they had just played their last game. He couldn't let the fire beat him or his team or his town. He was ready to fight. But, but, but . . . but hadn't the fire already won this first round? Wasn't everybody already knocked out? Shouldn't they all take a seat and live to fight another day? Prinz finally cornered his brother-in-law Scott in the garage of the crowded home—most of the family was staying there—and asked for help.

"I just don't know what to do," he lamented.

He thought Scott would entertain his thoughts and engage in a discussion. But his brother-in-law, a plainspoken sort, left no room for debate. His answer was as obvious as the smoke still billowing down from the hill.

"Yes, you do know what to do," said Scott in a scolding tone. "You can't ask these coaches to coach. You can't ask these kids to play. Forget about how awesome it would be. You can't ask people to play a football game when they don't know how they're going to live their lives."

Prinz knew Scott was right. So at 2:47 p.m. on Friday, just when the team should have been gathering at Paradise's Om Wraith Field to battle Red Bluff, with the players still expecting to play somebody some-

time and keep their title hopes alive, Prinz typed the toughest message
of his coaching career, directed to his team and school officials.

> It is an extremely difficult and emotional decision, but it is
> best to not continue with our season. Many of our players and
> coaches have lost their homes, possessions, and are devas-
> tated by this fire.
>
> We appreciate the support and concern of other football pro-
> grams in the area, especially Red Bluff. We are so grateful to
> our student/athletes and coaches for an incredible season.
>
> I would love to have a storybook ending, but the reality is we
> need to focus on our families and rebuilding our lives.
>
> I love and respect all of you. We will get together soon to cel-
> ebrate each other and our season.

It is a good thing his team couldn't see him when he pressed Send,
because Coach was crying. His town had been leveled, his football team
was in ashes, their championship-hopeful football season had been
gutted, he and had no idea whether they would ever play again or if he
would ever coach again.

Rick Prinz's life was out of gas, and he didn't know how he was
going to keep pushing.

TWO

"Dad, I don't want to die, but I'm gonna die."

NOVEMBER 8, 2018

What is dark within me, illuminate.

The kid was stranded at the top of a raging mountain. He was going to be burned alive and alone. He was going to die without his family, without his friends, with only his panting dog Gracie by his side. He was going to die in a scorching parking lot surrounded by a roaring fire and screaming strangers.

Spencer Kiefer, just months after his sixteenth birthday, grabbed his impossibly hot phone, prayed it could hold a signal, and called his father, Greg, who was trapped in his own car while pulling the family trailer several blocks away trying to get his son. The teen began speaking and couldn't stop sobbing.

"Dad, I'm gonna die," he said.

Spencer was a middle linebacker for the Paradise Bobcats. He could dish it out. Earlier in the season, he had separated his shoulder simply from hitting the guy carrying the ball. He could play with pain. But he couldn't play with this. The smoke filled his lungs as he uttered what he thought might be his last words.

"Dad, I don't want to die, but I'm gonna die," he said. "Good-bye."

"Wait! Wait! You're not gonna—!" screamed his crying father into a phone that suddenly went dead.

Spencer's mother had felt it coming. Earlier on the morning of November 8, she knew something was terribly wrong. Shelly Kiefer, a first-grade teacher at Paradise Elementary, kept waking up during the night. It was the stained glass butterfly wind chimes, four sets, hanging on the back porch. They kept rattling, a noisy rattling, which meant the wind was really blowing. She finally fell asleep, woke up late, jumped in the car for the quick drive to school, and saw a huge plume of smoke. She stopped. She turned around. She drove back home to warn husband Greg and her two children.

"Something's happening," she said. "If you get evacuated, go down to Grandpa's in Durham." The small town of Durham was about twenty minutes away, just south of Chico.

Shelly hurried back to her car and then off to the school. The kids would be there. If this smoke turned into anything serious, they would need her. As she drove, the smoke grew thicker but, strangely, she didn't see any fire. Nor did she hear any sirens. She pulled into Paradise Elementary at the same time one parent who was married to a policeman dropped off her child. There was nothing to worry about, the woman said assuredly. Suzanne sighed. But then five minutes later, the mother came back to pick up her child and was shouting for other parents to take their kids and get out.

Shelly couldn't leave. She had to stay with her pupils until they were all picked up. The school's halls filled with smoke. Parents were having trouble making their way to the school. Outside, Skyway Road was already gridlocked. People were driving on sidewalks or abandoning their cars. When the last child departed the building, she ran to her car and phoned her husband. Greg was preparing to evacuate. He was alone. Their daughter, Sarah, had left home earlier to head down the hill with her grandmother.

"Don't even try to get back to the house; you'll never make it," Greg told her. "Just go down the hill yourself."

"Where's Spencer?" Shelly asked.

"He's already gone to school," Greg said.

"By himself?"

Spencer had calmly shrugged at the warnings and left the house like it was any other day. He was going to be fine. He could drive through anything. He was still a kid, but he was becoming a man. He had just earned his driver's license. His father had given him an aging Ford F-150 pickup truck as a present. He had not driven it many places. Growing up on a quiet street in Paradise, there weren't many places to go. But he still knew how to drive, and he knew the directions to the high school, and what else mattered?

Spencer liked driving to school. He liked rolling that big truck into the parking lot and pulling up next to his teammates and hanging out before class. After school, he would return to the Ford, put on his uniform next to the truck bed, then hang out there until it was time to run down to Om Wraith Field and knock the snot out of someone. The truck was cool. The truck was safety.

So Spencer wasn't that worried until, while sitting with friends on his truck in the parking lot on the morning of November 8, dark blotches began falling from the sky. They looked like coal-colored leaves. They smelled like burning trash. They were giant pieces of ash. Something big was on fire. Then principal Lauren Lighthall came out to tell everyone to go home.

"Oh, shit," Spencer said.

He immediately jumped in the truck and raced home to his father. Greg was recovering from back surgery and would need some help. Spencer drove against traffic, trying to ignore the gridlock headed in the other direction, and finally pulled into their ash-filled driveway. He ran inside and grabbed a trash bag, filled it with photos and medicine, threw it in the truck, then helped his father hook up a trailer to Greg's truck. All around them was smoke. He could see flames in the distance. This was going to be bad.

They had lived in their home seventeen years. Three bedrooms, two and a half baths, on a half acre, with a wraparound porch. It was their pride. It was their life.

They'd found it in a newspaper ad. Greg walked up and knocked on the door in a pouring rainstorm. The nice woman who answered invited them inside. They saw the hardwood floors and were sold. They called the Realtor and bought it on the spot at full price. They signed the bank papers that night. The time stamp was ten thirty. Even the address was lovely: 1678 Aspen Lane.

The Kiefers painted all the rooms, refurbished everything, new hot water heater, new air-conditioning, and the dream house slowly became a dream home. They raised two children there. The kids had ridden Big Wheels on the hardwood floors and finger-painted on the walls. The family room had a fireplace with a gas stove, which everyone would gather around for holidays, when they weren't chasing down Easter eggs in the front yard. There were no such hunts in the backyard, because, like many similar homes in Paradise, the yard belonged to the chickens and goats.

The Aspen Lane house not only held memories but also created them. There was the time Sarah became stuck in the bathroom, forcing Greg to take the door off the hinges and eventually replace every door handle in the house. Then there was the time Spencer's head became stuck between two staircase railings, and Greg had to remove one of the poles to quiet the crying child. It was the kind of house where the concrete walk in the front was embossed with children's handprints.

GREG LOVED THAT HOUSE. AS the smoke kept getting closer, he made the decision to send Spencer down the mountain with the family dog while he stayed put, watching for looters and maybe even helping the firefighters.

"I'll be fine, Dad," said Spencer. "I'll just drive down the hill."

He soon realized it wasn't going to be that easy. He took his usual shortcut to Skyway, but suddenly he found himself stuck on a narrow road filled with cars, with flames seemingly threatening from all sides. A policeman approached him with an order.

"Leave your vehicle, walk to the Optimo parking lot. Do it now!" the cop said, referring to a local Chinese restaurant landmark. "It's your only chance."

Suddenly it all became real. The truck couldn't save him. His cool new driver's license couldn't save him. The familiar hill out of town couldn't save him. He phoned his father, who had started navigating a different path down the mountain. For the first time, Spencer began crying.

"I've lost the truck!" he said. "Dad, I lost the truck!"

"Let the truck go," said Greg. "If you have to, let the dog go, too! Save yourself, Son. Save yourself."

As the son was making a life-or-death decision, so was the father. Earlier, when Greg was stomping on fires in front of his house, a fire chief had pulled up and ordered him to go. But the moment Greg heard Spencer was stuck, he turned the trailer around and headed into the traffic to find his son. He couldn't go. He wouldn't go. He would not leave the mountain until Spencer left the mountain.

Greg drove through streets that looked like they were lined with torches and pulled up to a spot where the fire was behind him and his son was somewhere in front of him. Either he was going to get his son or perish in the flames. Greg was a former US Army platoon sergeant who'd fought in the 1991 offensive against Iraq, Operation Desert Storm. He had been on the front lines. He had been bombed. But with the wind blowing flames at his back and his son's life dangling just a half mile ahead, he had never been in a position quite like this one. All he could do was wait, and pray, and try like hell to breathe.

"I'm going to be burned alive," Greg thought to himself before shouting into the phone, "I'm not leaving you, Son!"

"I'm really scared, Dad!" Spencer shouted back.

"I'm coming to get you!" But Greg couldn't go anywhere.

With his father still stuck about a half mile away, Spencer was able to maneuver his truck to the side of the road, then walked with Gracie, his beloved Labrador, for a mile and a half to a holding cell of horrors. There was fire in front of him and fire behind him. As the sky blackened and the smoke rolled, he arrived at the Optimo parking lot to join dozens of people who were crying or gasping or praying. They couldn't move forward. They couldn't turn around. They were the definition of trapped.

It was here that Spencer made the call to his father, predicting his death and bidding farewell. His father's offer of encouragement was silenced by the dying phone. Down the mountain, where Suzanne had already driven, she couldn't get ahold of either her husband or her son, and she spoke the awful truth into the smoke.

"If I lose Spencer, I'm gonna lose Greg," she said. "If one doesn't make it out, the other one won't make it out. Dear God, please save them."

For several hours, Spencer stood in that parking lot. Not once was he certain he was going to survive. Not even when firefighters busted out storefront windows so that the stranded could get inside away from the smoke. Not even when helicopters dropped water on the fire around them. The roar of the hot wind and the thickness of the smoke were accentuated by the *bang-bang-bang!* of explosions. He was standing in front of a propane store, and the tanks were exploding around him. He had begun the day with five bottles of water, but, after sharing them with Gracie, he was nearly out. By now, he was also seriously worried about his asthma. He stripped off his sweatpants and placed them over his mouth. "Breathe, damn it," he whispered to himself as he sucked air through the cotton. "Breathe!"

Nearly out of hydration and hope, Spencer walked out of the shattered storefront and clung to a group of policemen, figuring they would be the last ones to perish. Turned out that they were the first ones to offer a sliver of hope. Late in the afternoon, an officer announced that bulldozers had cleared the burning and abandoned cars off Skyway. At the time, another policeman who knew the family had taken Spencer's keys and found his astoundingly unscarred truck and drove it up to him.

"Get in," the cops said. "You're driving in a caravan down the hill."

"Hell yes I am!" exclaimed Spencer.

So the sixteen-year-old with three months' driving experience calmly steered his truck fifteen miles downhill through flames. He thought how this was no different from a summer tackling drill in hundred-degree heat with coach Paul Orlando exhorting him to keep pounding. This was no tougher than one of Rick Prinz's three-hour August workouts that end in gasping and sweat. If he could survive Paradise football, he thought, he could do this.

Spencer remembered his mother's directive and eventually steered off the Skyway and headed south toward Durham, all the time charging his phone. At 5:53 p.m., nearly ten hours after his nightmare began, he pulled into a parking lot in Durham and slumped over the wheel. There were burn holes in his clothes. His sweatpants had turned black. His dad was still on the hill, and he couldn't reach him. He knew he needed to call his mom, but he knew he would cry, and first he wanted to celebrate. So he called a buddy, Trevor Rickson.

"I'm alive, bitch!" he screamed.

When Greg found out his son was safe, he could finally leave the mountain. Tugging a twenty-eight-foot trailer, he navigated a one-lane back road down around the fire, and went straight to a Costco parking lot, where, at about nine at night, father and son collapsed together in tears. Their nightmare, however, was just beginning.

That night, while Spencer slept on the floor of a family friend's house, Greg dozed in the front seat of his truck. He was exhausted and in shock.

"What happened to our life?" he kept saying over and over.

Two days later, Greg was still in the front seat of the truck when someone from the neighborhood approached and tapped on the door with a question:

"Are you okay?"

Greg just rolled up the window and turned away.

A buddy from the Paradise Sheriff's Department stopped by later with a video showing that the Kiefers' dream house had burned down to the stubs.

"Nothing is ever going to be the same," Greg said.

This included Spencer.

THREE

"There's no way I'm walking out now."

WINTER 2018

*These raging fires will slacken, if his breath
stir not their flames. Our purer essence then
will overcome their noxious vapor.*

The town of Paradise was named either by a weary forester or after a wild saloon. Each story works.

The land in the Sierra Nevada foothills that is now called Paradise was inhabited for perhaps ten thousand years by the Maidu Indians and settled later by prospectors seeking gold in the mid-1850s. As recounted by Robert Colby in his book *Images of America: Paradise*, neither legend has ever been proven, but both make sense.

In one, sawmill owner William "Uncle Billy" Leonard named the town in the summer of 1864 after driving a wagon train up from the sweltering Sacramento Valley to the cool shade of the heavily forested ridge.

"Boys, this has got to be Paradise," he told his crew.

In the other story, the town was named after a local watering hole called the Pair 'O Dice Saloon, and, indeed, an early map lists the city as "Paradice."

However it was named, the first post office was established there in 1877, and both stories proved prescient. Paradise is both lovely and wild. Its earliest residents had moved there to cash in on the California gold rush of 1848–55, only to be disappointed that Paradise was one area of the Sierra Nevada foothills that did not possess gold. It was surrounded by mines yet was no fountain of riches itself. Those first settlers were a hardscrabble lot that eked out a living cutting timber, working in sawmills, and farming fruit orchards.

Those who lived in Butte County's surrounding valley communities viewed Paradise as simply a place one traveled through to access the working gold mines and the rich timber farther up the hill. They didn't understand how anybody could actually live there. They even sometimes referred to Paradise as "Poverty Ridge." But its residents loved the place for its fresh air, its splendid surroundings, its remoteness.

And, eventually, its football.

"Tough lives, tough game, it fit us perfectly," said Greg Bolin, a former Paradise mayor who grew up on the scraggly Bobcats field.

The first recorded football game between Paradise schoolkids and outsiders took place in 1952, when a group of local eighth and ninth graders, coached by a physical education teacher named Joseph Kearney, played a group of sophomores and juniors from nearby Live Oak. The Paradise kids were walloped, 47–0, but merely by showing up, they demonstrated that they were willing to be Davids fighting off Goliaths.

"If you play for Paradise, you won't get away with being babied," said Bolin.

The first high school season was 1956, when the team went 5-3-1. It had only one winning season in the next seven years, then finally broke through with its first league co-championship in 1964 with a 7-2 record under the venerable coach Om Wraith. The Bobcats would win four more titles in the next eleven years, as the ancient cement stands lining one side of the weed-choked football field filled up, and the tough Paradise culture was established.

"Football became the thing that bound the town together," explained Jay Bell, athletic director at the high school from 1985 to 2004. "Paradise is a different kind of town, and we had a different kind of football program."

The program was epitomized by the words of the coach for whom the school's field was eventually named.

"You know where to find *sympathy* in the dictionary?" Wraith used to ask his team during grueling three-hour practices. "Between *shit* and *sorry*." Needless to say, his football skills were more precise than alphabetizing skills, but the players got the point.

Bolin, whose family moved to Paradise from Los Angeles in 1967, partially because of his brother Tim's asthma, had a front-row seat for the formation of this culture.

"The air was clean, the water was fresh, there was room to run. It was the perfect escape from the city," he remembered.

As he soon learned, it was also the perfect place for all sorts of free spirits. What was once a drive-through town morphed into its own sort of hidden destination filled with spirited characters resembling the pioneers of years passed.

"Some people came up here to get away from it all, others came up here to hide," he continued. "It's a giant a cul-de-sac. You had to want to be in Paradise to be in Paradise. It wasn't a drive-through town. It was a tight-knit group, everybody knew everybody, people knew everything you did, but they also had your back. It was a mountain town that was different from those nearby valley towns of Chico or Oroville, but we embraced that difference."

The community shared a certain toughness that came from the rural environment combined with a lack of local industry. It wasn't a rich town. Before the fire, many of its twenty-eight thousand residents were retirees or laborers who would leave the ridge during the day to work in the neighboring valley cities.

"Tough lives, calloused hands, kids would chop wood, do chores, work construction with their fathers, a lot of broken homes, not a lot of distractions, just tough lives," said Bolin.

One of the more visible differences between Paradise and its more affluent neighbors, born of the town's unvarnished beginnings, was football. Bolin was nine when he arrived. He joined his first youth football team a year later and learned quickly that "football in this town is brutal. I wanted to play," he recalled, "and a lot of kids wanted to play, because we could hit somebody and not get in trouble. And, man, we hit people."

In one of his youth team's first games, they were so rough, officials threatened to throw them out of the league.

According to Bolin, "We were smash-mouth, grit, blood. You wanted water, you were a sissy. You had a complaint, just shut up and get back in the huddle."

The junior football ethos came from the culture of the high school squad, whose players were town celebrities, such was their weekly popularity. They hit as if releasing their town's frustrations. They fought in the face of their town's perceived slights. The Bobcats were exactly as their town viewed itself: scrappy underdogs who never quit.

"The football team took their inspiration from the town, and the town took its inspiration from the team," observed Bolin. "It was pure storybook. As a ten-year-old boy, watching them play in front of all those people, bonding like they were, I knew that's where I was going to have to be on Friday nights."

The Paradise High Bobcats' aura was helped by their field. It wasn't pretty. As the season progressed, whatever grass existed disappeared under the constant ridge rains. Visiting schools, some of whom played on shiny new turf, would step onto Om Wraith Field and freak out.

"Mud was our thing," said Bolin. "We knew how to play in the mud. Paradise people didn't mind getting dirty."

Their reputation for being tough mudslingers became so prevalent, it was difficult to convince opponents to play there. In fact, one of the everlasting team mottos became "Don't Come to the Mountain."

"Every year, because of our weather, our field becomes almost a disaster area," said Jay Bell. "Teams don't want to play in the swamp. Teams don't want to have to come to Paradise. Our coaches turned that to our advantage."

Bolin's teams won sectional titles in 1973 and 1975 in games that featured muddy scrums, postgame brawls, and one occasion when the players had to restrain then-coach Art Guerra from fighting an Oroville coach whose team had torn the helmet off one of his players.

Bolin graduated and eventually became the owner of a successful construction business. But the Paradise football program he left behind fell into disrepair. The Bobcats became better known for fighting than for winning football games. From 1981 through 1997, as Om Wraith eventually retired, and the program lost focus, they had just one winning season.

Then, after a 1998 season in which they went 2-8, Bell noticed this young Paradise assistant coach who was also a local youth minister. He was tough, but embracing, and he coached like he was preaching, and the connection between him and the players was clear.

"He just cared for the kids," remembered Bell. "They'd be in talking to him all the time, hanging around his office between classes. Kids just gravitated to him."

Paradise thus hired Rick Prinz as head coach to replace John Luciano, and the winning returned. The Bobcats made the playoffs in his first season, they were section runners-up his second season, and then they became unbeaten section champions his fifth season. It was the start of twenty years of historic success. It was so contagious, even a former player named Greg Bolin joined the fun as a member of the sideline officiating chain gang, a job he still held through 2021.

The team was on a roll. The town was in its thrall. Prinz had it all.

Yet there he was, on November 10, 2018, flat on his back, his future seemingly in ashes.

Prinz was getting up from an air mattress in the living room of his brother and sister-in-law's house when he received the first text of hope.

It was a Saturday morning, two days since he'd fled down the mountain from the fire that had gutted his community. He had not slept. He had barely eaten. The coach had spent some of his time on the phone soothing some of the many players whose families had lost their homes. He had spent the rest of his time mourning with his six coaches, who'd lost everything. He had spent two days thinking about others, but this text brought him back to himself and the immediate question surrounding his own life and impending retirement. What about his house? Had it burned? Was there anything left?

A former player who was an electrician sent the text from the off-limits remains of Paradise. It was a photo. It was fuzzy but beautiful. It was the Prinzes' cedar home, surrounded by rubble but still mostly intact.

"Honey, you gotta see this!" he shouted to Veronica.

His wife ran to the phone, clutched it in her hands, and fell to her knees in tears.

"It must be a sign," she said, sobbing.

But of what? He still didn't know if the house was inhabitable. He couldn't tell how much of his property had burned. One distorted, darkened picture didn't convince him that he should return to the team. It didn't even mean he and Veronica should return to the town. It didn't mean anything other than this: their lives hadn't been *completely* shattered. The day before, Rick had said to Seth, "If our house is still there, that must mean God wants me to stay." But, in reality, even the knowledge that the family's home had survived the Camp Fire didn't give Prinz much direction about where to go next.

Two days later, that direction began to take focus following a three-hour trip on a rickety school bus.

THE SAN FRANCISCO 49ERS FOOTBALL team was looking for a way to connect with the victims of the fire, and so the club invited the displaced Paradise football team to attend its Monday-night home game against the New York Giants. The Bobcats gathered at the Pleasant Valley High School parking lot in Chico on the morning of the game. It was the first time they had been together as a team since the previous week, before the fire. This was, finally, real proof of life continuing.

Prinz saw his kids, and his throat thickened. Between tears, he noticed they were also crying. There were deep hugs. There were convoluted tales of escape. A local newspaper reporter wanted to interview Prinz, but the coach couldn't handle it, shaking his head at a question that was repeated three times. Through it all, there was shivering, lots of shivering. The temperature was in the low forties, and most of the kids didn't have coats. They didn't forget them; they simply didn't have them anymore. Their winter clothes had burned up in the fire.

For not the first time, athletic director Anne Stearns swooped to the rescue. Out of nowhere, a truck pulled up with a box of mismatched Paradise sweatshirts that she had pilfered from the student store earlier that morning. She had been the first faculty member to return to campus since the fire. She talked her way past the police, drove around the embers, sneaked onto the campus, grabbed the shirts, and ran. She figured somebody would want her to eventually pay for the merchandise. They could send her a bill.

After she pulled into the parking lot with the shirts, the players gratefully pulled them on, trudged onto the bus, and with few exceptions, fell asleep against their seatmate's shoulder. Some three hours later, when the bus pulled up at Levi's Stadium, a reporter asked Prinz if the kids were fired up.

Prinz gestured toward the back of the bus and shook his head.

"No, they're not fired up; they're all sleeping," he said. "For most of them, it's the first time they've slept in four days."

After the bus doors opened, the thirty-five players, sixteen cheerleaders, and eight coaches were awakened in a flash. Literally. Waiting for them were dozens of cameras. They initially thought the media was there for some attending celebrities. Then it hit them: *they* were the celebrities. The cameras and reporters jostled for position outside the bus. They lined up to talk to the players. It was then, for the first time, that Prinz realized this local nightmare was actually a national story. He had no idea. None of them had any idea.

"Hey, Coach, can we talk to one of the kids who lost their homes?" asked a reporter.

"Pick a kid, any kid," said Prinz. "They pretty much all lost their homes."

The Bobcats players gathered in the stadium tunnel, where they were met by 49ers general manager John Lynch, a Hall of Fame safety who noticed that center Trevor Curtis was wearing an Oakland Raiders cap.

"How dare you wear that in here!" he shouted in a joking exchange that was picked up by local television microphones.

A couple of weeks later, a group of Raiders fans contacted Prinz and personally gave him a box full of Raiders gear for Curtis. The Paradise story had that sort of perverse power. Prinz knew it the moment the busses pulled up in front of the media. And he knew it when this group of kids in mismatched Paradise jerseys and sweatshirts walked out of the tunnel and around the stadium before the national anthem. The fans roared, not for the 49ers, but for them. "Para-dise! Para-dise!" they screamed. As the chanting slowly stopped, the shouts of encouragement from the stands began. "You guys hang in there! You guys can do it!" Soon the smallish players were standing arm in arm with the giant 49ers players for "The Star-Spangled Banner," feeling warmth for the first time in days, a football team again.

Although the 49ers lost a heartbreaker, 27–23, the Paradise Bobcats didn't care. They knew loss, and this wasn't it. This was relief. This was escape. They came down from their end zone seats after the game and walked across the field to appear on the 49ers' postgame show, where they were met by a tearful Jeff Garcia, a retired quarterback who'd been the franchise's field general in the early 2000s.

"I'm one of you guys," he told them. "Came through a small town and a tough kid just like you guys. Hey, this adversity, this struggle you're going through, is a difficult time in your life, but you're going to battle through it and come out on top . . . rally the troops and rally the town, right? Rally the town!"

Rally the town. They had never thought of something like that until he said it. Rally the town? But didn't they need a place to sleep first? They left the postgame show and began walking back across the field toward the busses when something happened that was totally amazing yet completely appropriate. Somebody found a football. Then somebody else had an idea.

"Coach, what do you think?" asked one of the kids.

Prinz just smiled, and soon they were lining up and running plays, on a real NFL field, with an offense and defense and their coach directing them. Though clad in jeans and sweatshirts, goodness, they were running real plays! For a moment, everything came back; everything felt right. Then former 49ers great Steve Young sauntered past after finishing his ESPN postgame duties as an analyst.

"Are you the Paradise team?" he asked. Young, San Francisco's starting quarterback throughout the 1990s, led the NFL in passing percentage five out of six seasons from 1992 through 1997. He spent the next thirty minutes talking to the boys, encouraging them. Young also signed the Paradise jerseys of the lucky few who had saved them.

"Hey, I'm picking up those jerseys next week like I do at the end of every season," Prinz said with a smile. He didn't mean it, of course. Anybody who grabbed his jersey or had his jersey wadded up in his trunk or was actually wearing his jersey while escaping from the fire could keep them, now and forever.

But by the time they reached those old yellow school busses, the glow was fading. The kids dropped into the cheap seats, and once again they slept. About halfway home, the bus pulled into a truck stop for a bathroom break. The place only had one working bathroom. The cheerleaders and the older folks lined up. The players hustled off to the bushes. Standing in that long line in the middle of the night, Prinz heard something that ended this night of fantasy and returned it to reality.

"Shhhh," said one cheerleader to another in the conspiratorial whisper of someone who had just figured out a way to rig the lottery. "I hear you can go to the thrift store and get a whole bag of clothes for five bucks."

The other cheerleader responded, "Oh, man, I need to go tomorrow."

Standing behind them, Prinz's heart sank. He had heard kids at his hardscrabble school talk about poverty before. He had heard about their hunger pangs and felt their chill, and had usually been able to do something about it. A few bucks for lunch. Arranging a ride home. He was always the adult in the room, always the savior, often able to swoop in and fix his students' problems. Not now. Not anymore. How could he supply these kids with clothes when he didn't know if any of his own clothes had survived the smoke damage? How could he give them money when he wasn't sure whether he was still going to have a job?

This was a low point. But it was going to get worse. In those first few months after the fire, the low points always got worse.

FIRST, THERE WAS PRINZ'S LIVING situation. His house was officially intact, but he had no idea how much of it actually remained. It was a month before he was even allowed to return to the city, and that was only because he agreed to an interview with ABC News correspondent Kayna Whitworth. Rick didn't want to relive the pain so soon, but he wanted to see his house and his town, and this was the only way. Whitworth had permission to take him to Om Wraith Field for the interview, where he had his first look at the high school. Driving up the hill felt like driving through a war movie. Rubble everywhere. Still-smoldering trees on all sides. Military police at every corner. Amazingly, the high school was still standing, but everything around it was burned, from surrounding trees to an upper building that housed the team's film room.

He noticed that a giant outlying steel storage of their equipment was still standing. He walked over, looked inside, and sighed. During the fire, the container became an oven. The pads and helmets and blocking bags had been reduced to a few inches of ash. A pile of wood chips still smoking.

"This feels real now," he muttered to Whitworth. "Let's get out of here."

From there they drove up to his Coldren Road home, where he was stunned again. His actual home was still standing, but the homestead was gone. His shop was gone. His barn. His deck. His landscaping? In ashes. A stump was completely burned just inches from the house. Amazingly, Seth's old truck—the one that Prinz insisted he abandon— still sat in the driveway, though shrouded in layers of ash.

Prinz ran inside the house to quickly grab some computers and clothes before the police showed up and escorted them out. He didn't need much coaxing. The smell was horrific, like being on the edges of a campfire, and he just wanted to leave. But he knew he had to come back. This was where he lived. This was still his home.

A few days later, with the sort of police escort given all the returning residents, Rick and Veronica returned. They had to wear Hazmat suits. They were given an hour to sort through their belongings. They took photos and electronics and linens. Most everything else would have to be thrown in a dumpster, including every piece of furniture, because they stunk of smoke and were badly smoke damaged. The walls. The carpets. The drapes. The couches. The kitchen table. The antique nightstand. All two thousand square feet, which had embraced this family for twenty years, would have to be cleansed—or else destroyed.

"Oh my gosh, can we still live here?" Prinz asked his wife.

"Do we have a choice?" she replied.

For the first two weeks after the fire, Rick and Veronica thought they had an agreement to rent a home in Chico temporarily. But at the end of the second week, the owners decided to sell. Of course they did. Out of options, the Prinzes returned to their hometown of Willows, a tiny farming community about an hour southwest of Paradise. There they occupied a vacant home owned by a close friend, Jeff Fleming. It had three bedrooms, two bathrooms, but no bed. And they weren't going to buy a bed, because that would be a concession that they were setting down roots here, and that could never happen. So, for five months, they slept on air mattresses and awoke with sore backs and high tension.

Prinz wanted to move back home even before the water supply had been declared safe; his wife wanted to wait. They fought about it. Veronica didn't want to live in ashes. And, as usual, she was right. The place on Coldren Road was simply not in livable condition. Every Saturday and Sunday, they would drive up to the house and clean. They would dump burned tools, move fallen lumber, empty out the house, and throw everything in the bed of a truck and haul it to the dump. Again and again and again. Five loads every Saturday. The trashing of their lives. They were doing it all with no water, shaky power, and no federal help. Because the Prinzes' house did not burn, they did not qualify for federal aid. Their homeowners' insurance offered some help, but only with cleanup, lost property, and landscaping. It didn't come close to making them whole. It was a different sort of scorching.

"Sometimes I wonder if we would have been better off losing the house entirely," Prinz once said to his wife.

"Don't say that," she replied.

Amid the chaos, in early December, school started again, only it was strictly an online school held at the Chico Mall.

"My one rule in life is never go to a mall," Prinz told friends. "Now I'm going there every day."

Working out of a former LensCrafters store, the school operated like a drop-in lab. Kids would show up, work on computers, and the teachers would monitor them. There wasn't much anybody could do but stay busy. It was four hours of daily drudgery, but at least some kids were showing up, and Prinz realized this was his chance. This was where he would hold his first team meeting for the 2019 season. This is where he would . . . what, exactly? He couldn't say. It's tough to have a team meeting when you're not even sure you're going to field a team. It was odd to start coaching when he didn't know if he would even be their coach. But he saw a few of his players, wondered what they were thinking, and figured he might as well pull them together. He tasked Lukas Hartley, his rising senior running back and team leader, to spread the word. It was just another odd announcement in a season that would be filled with them:

"Tuesday Dec. 18, 9 a.m., team meeting, Chico Mall food court."

The mall wasn't open that early, so they would have the place to themselves. When Tuesday morning came, Prinz went to the empty food court tables surrounded by a pizza franchise and Chinese takeout and burger joint, and he waited. And waited. And waited. He had no idea how many kids would make it, because their classes were online, and few of them were coming to the drop-in labs.

Eventually Hartley showed up. And running back John "J. D."

Webster. Then another running back. And another one. The meeting began with just twelve team members present. Prinz looked around the room and winced. He knew it would be bad, but he didn't know it would be this bad. From 104 kids to 12, his program had officially been decimated. Prinz was thrilled to see them but devastated to see so few. Everyone started talking, exchanging stories about sleeping on couches and running from flames, and, for a moment, it felt like a family reunion. But the boys quicky turned their focus to the one football question their coach could not answer. They wanted to know about the future. When that happened, this famously honest and plainspoken coach knew what he had to do.

He lied.

"We want to play next year," said Hartley.

"We'll play next year," Prinz assured him.

"We want our team together again," Lucas said insistently.

"We'll get 'em all back together again," said Prinz.

The players ended the meeting just before the morning shopping rush overtook the mall. They were hugging and feeling good. Prinz, however, was in a quandary. What did he just do? Why did he assure the players of something that might not happen? Why couldn't he tell them he was going to retire? Maybe, in a way, this meeting was telling him that he *shouldn't* think about retirement? Prinz wondered. That night, he fell to his knees by his air mattress and prayed, "God, give me a sign, make something happen, bless me with something that will help me make my decision."

A few days later, Rick was invited to a meeting of school district officials. He didn't want to attend; he was exhausted simply keeping his life and job afloat. But Anne Stearns convinced him he should be there.

In the middle of the meeting, Stearns walked up, set a smartphone in front of him, and San Francisco 49ers head coach Kyle Shanahan appeared on the screen. The 49ers were on the last legs of an awful 4-12 season, their second under Shanahan, but the coach was bringing hope.

"I want to reach out to you and let you know we appreciate you," Shanahan said to Prinz. "I hope we finish as strong as you this year; hopefully we'll do that for you next year." (Indeed, in 2019 San Francisco would go 13-3 and make it all the way to the Super Bowl.)

Shanahan continued: "It's unbelievable what you had to go through. I know you guys are trying to persevere. We have a weekend set up for you if you want it, with a couple of tickets to go to the Super Bowl."

Wow. Prinz was blown away. Shanahan had just given him two tickets to Super Bowl LIII at Atlanta's Mercedes-Benz Stadium. Somebody noticed. Somebody cared. Somebody understood. In February, he and Veronica sat among the seventy thousand other football fans to watch the New England Patriots bust open a 3–3 deadlock in the fourth quarter to fend off the Los Angeles Rams, 13–3.

By then, Prinz had received yet another sign, again courtesy of the 49ers.

The organization had invited the Paradise Bobcats to attend a December 30 season finale watch party in their honor at the rival Pleasant Valley High School in Chico. A TV crew would be there to interview the team members on the pregame and postgame shows and they would watch the game together in the school's auditorium. It would be a fun distraction. Prinz encouraged his team to attend, and fifteen kids showed up in their now-usual mismatched jerseys and borrowed winter coats. It was going to be a neat night. Then it wasn't.

While they were standing outside waiting to enter, they were suddenly stunned by the sight of the state champion Pleasant Valley Vikings football team walking in together. Pleasant Valley was the high school of the Green Bay Packers' star quarterback Aaron Rodgers. It was a well-funded program with more players, bigger players, and much more exposure. The Vikings overshadowed the Bobcats in terms of public perception, and now, even during what was supposed to be a night for the hurting Paradise fire victims, giant Pleasant Valley once again stole the show.

Seeing the champions both saddened and angered the Bobcats. They could have also been champions of a different division. That swaggering pride could have also been theirs. The divergent paths taken by both teams since the Camp Fire blaze was starkly clear. Why would any TV crew want to highlight the situation? Why would the Paradise team want to appear on television looking like raggedy orphans next to their accomplished gridiron rivals, who were all wearing their crisp, clean jerseys?

"Fuck this," said one of the Bobcats.

"Yeah, fuck this, I'm out of here," muttered another.

Three players left. As for junior center Blake White, he simply refused to get out of his car. Everyone complained. They looked at Prinz. What should they do?

"You don't want to go in there, don't go," their coach said. "Do what you feel."

Most of them eventually went inside. After all, it was a free meal. But they all felt horrible.

"This is salt in our wound," muttered Prinz, but he initially played the part of the dutiful coach. He walked inside, talked on television, and settled in to watch the game. Then it hit him.

"Forget about doing the proper thing; what am I doing here?" he asked himself. "These are nice people, but they don't know. They just don't know. They don't know how this is affecting us. They don't understand what it's caused for us emotionally. I left my brother and sons who were painting my smoke-damaged walls for this?"

Amid days of low points, this was yet again one of the lowest.

"This knocks me on my butt," Rick said to no one in particular, but everybody heard him.

Prinz was supposed to stay and watch the entire game and eat the food and enjoy the camaraderie. Afterward, he was scheduled to appear on the postgame show. He decided he couldn't do any of it. In the middle of the game, he blew off the hospitality and the postgame show. He left to drive back to Paradise, so that he could help everyone paint his house while there was still daylight.

But before he did, as Prinz was heading out the door, he looked at his kids huddled up on one side of the room, their eyes heavy, their jeans dirty, their lives a mess.

And Rick Prinz made a decision.

He pulled out his smartphone and called Veronica.

"There's no way I'm leaving these kids in this condition; there's no way I'm walking out now, not on this school, not on this town," he told her. "If we still have a team, and I still have a job, I'm going to coach next year."

Veronica chuckled softly. She knew already. In fact, she had known for more than a month, from the moment she saw her husband crying when he canceled the season. She knew he couldn't leave his young players like this. She knew he couldn't leave the town like this.

"So what happens now?" she asked.

"I have no idea."

FOUR

Heavy Lifting

SPRING 2019

*Our torments also may in length of
time become our elements.*

"Anybody have a football?"

Rick Prinz looked around—*clang!*—at his new surroundings—
"Heads up!"—and frowned. He was sitting in a corner of a giant cage in
an industrial warehouse next to Chico Airport. This was where Paradise
High School officially moved in January 2019, two months after the fire
decimated the town. This was where they would hold classes for the rest
of the school year while the remains of their city were being cleaned up.
The high school campus itself was mostly untouched by the fire, but all
around it were ashes. The old building still lacked power and sanitary
water, so they would finish the year in an oversized warehouse.

Prinz stood up from his desk—*thud!*—and began slowly walking—
crash!—out of the cage. His office was a makeshift weight room cluttered with ancient barbells and weathered benches. He was assigned one
of three desks in the space, the rest of which was populated by sweaty
teenagers trying to do a few reps. Outside the cage, in an empty concrete
space, somebody had set up a basketball hoop, and it was surrounded by
sweaty teenagers shooting jumpers. In a far corner of the concrete area,
still other sweaty teenagers were playing floor hockey.

The coach's new working conditions were so loud, he had to shout to be heard. It was so cramped, he could barely move his desk chair without hitting a dumbbell. What's more, not only was the WiFi erratic but also the WiFi box was battered from being hit by stray basketballs. And, oh yeah, there was no air-conditioning in the garage, so he often worked in a pool of sweat.

Little did Rick Prinz know that on this mid-February day, his reality check was about to become even more real.

It was sixth period, which meant Prinz's PE class was essentially filled with football players lifting those weights. It was a small group; not even close to a full team. The last time he felt any semblance of a full squad had been a couple of weeks earlier, on January 18, when they held their 2018 team banquet at a local catering hall. The players hugged one another. They were happy. They cried together. They felt like a powerful team again. It didn't last.

Once school restarted in this warehouse, many of the teens drifted away or just plain disappeared. They couldn't take the ninety-minute drive from some distant town where they lived in some tiny mobile home. Some of them moved out of state. Some of them just couldn't get out of bed. The ones who did attend, they lived mostly within a three-hour radius, put up in a variety of accommodations—apartments, trailers, rental houses, hotels—that were paid through insurance or life savings. The players were everywhere, and nowhere.

As Prinz walked out of the cage on this February afternoon, he silently counted the number of varsity players present out of a roster that once numbered fifty kids.

He counted twenty-two. Exactly enough for one offense, and one defense, with no subs.

"I'm going to regret this, but it's time," he mumbled to himself as he navigated past the weights and the basketball hoop and the floor hockey and reached the giant entrance leading to an adjoining vacant lot. As he has done every year, he was going to start spring practice for the upcoming season, conditions be damned.

"Okay, everybody!" he announced. "Today our new season officially starts again. Today we're going to go outside to the field and play some football!"

He was lying again.

The "field" was a weed-choked lot filled with rocks and trash, food wrappers and soda cans, and giant holes next to piles of dirt. "Football" was also a misnomer, as most of the kids were wearing old blue jeans and casual shoes. Even if they had gear, there was no place to dress. The "locker room" was a tiny bathroom with two urinals, an open stall, and one sink. A makeshift sign on a nearby wall attempted to be cheery: "We're from Paradise, we couldn't be prouder. If you can't hear us, we'll yell a little louder."

Right about now, that school spirit had to be yelled *a lot* louder. Their first workout since the fire was filled with a bunch of kids with no sweats, no cleats, and no chance.

"Everybody out here!" shouted Prinz, stomping to the middle of the tangle of weeds and litter adjoining a giant dumpster. "It starts here, and it starts now!"

One of the returning players, junior center Blake White, was stunned by what he saw.

"This field is really bad: potholes, dog shit, rocks," he said. "But, whatever. If we can play football, maybe we'll feel normal."

Also showing up was junior running back and linebacker Jeff Trinchera, who was staying nearly two hours away in the small city of Wheatland. He was wearing jeans and a collared shirt and a perplexed expression.

"I forgot my clothes," he said. "It's too far to go home and get them. I'll play in my long pants."

Then there was junior lineman Jose Velasquez, who escaped the inferno with only the clothes he was wearing and a school backpack. He lived with five people in a two-bedroom house and couldn't sleep.

"I'm tired, so tired," he said. "But I need football."

Also present was junior defensive back Dylan Blood, expressionless, dazed, still haunted by his driving escape with Velasquez through the fire.

"It's crazy," he said. "It's still crazy. It should feel good. It doesn't."

Running back and safety J. D. Webster, who was living with his parents and sister in one room in Chico, said passionately, "We never got to finish off our season; that sticks with all of us. This year we're gonna show everyone what we should have done last year." While Webster was talking, two kids behind him were cursing out each other loudly over a basketball argument. Sometimes the noise in the warehouse was deafening. The field may have been a mess, but it was also a refuge.

Junior running back Stetson Morgan, who normally wore number 2, also came dressed in school clothes, but for a different reason. He lived with his parents and brother in a trailer on a piece of farmland forty-five minutes away. It was cramped. He slept some nights on a chair. He didn't have the energy to go home last night, so he'd slept over at a friend's and didn't have a change of clothes.

"I didn't go to my house last night, so I don't have my pants," he said plainly. "But this is why I'm in school: to play football."

One of the last to arrive was linebacker Spencer Kiefer, still not quite understanding how and why he was still alive.

"I don't think anybody understands what we've been through," he said bitterly. "Even the people in Chico don't understand. We're a nuisance. We clog up their traffic. They look at us, like, 'You're just a Paradise kid; you're not worth the trouble.'"

The Kiefers had moved into a rental home in a tidy but crammed Chico neighborhood where the houses sat on top of one another, and it slowly drove them crazy. There were no driveways, and sometimes the family would spend thirty minutes looking for parking. And when his father, Greg, did find a spot on the street, he would get notes on his dashboard claiming his truck was too big.

"I had a half acre, and now I have nothing," he lamented to his wife, Shelly.

They would wake up at night to hear water running, toilets flushing, deep and loud conversations from next-door neighbors. The country folk didn't do too well in the city. And their tough country son had a difficult time outrunning his personal escape story.

Spencer had nightmares for three months after the fire. He would awaken at three in the morning and come downstairs and wander around the kitchen. He would find his father sitting at the table, and neither of them would know why they were there. Greg had suffered bouts of posttraumatic stress disorder from fighting in Desert Storm. He knew PTSD when he saw it. And he was seeing it again—in his son. Almost overnight, Spencer went from drinking soda pop to slamming back beers.

"We're both trying to find a way to live," Greg said.

Spencer's way was to sometimes wear his gray Nike shirt with two burn holes in the back to school. Or he would wear that same pair of smoke-damaged sweatpants that he once held over his mouth to block out the acrid smoke.

"Hey, I still have these; I made it through," he would say.

Spencer had a blank look in his eyes sometimes, as if he were seeing the world through a new, jaded lens.

"He went from a sweet teenage boy to a young man who didn't give a crap about a whole lot," Greg said empathetically.

Spencer shrugged when he heard his father say this. Greg was right. But the teenager didn't care.

"I didn't need any help. I got over this on my own," he insisted. "As soon as the fire was over, it was like, 'I'm invincible now. If the fire can't get me, nobody can get me.'"

It was this attitude that Spencer brought to the field four months later on the first day of practice. From the outset, it was clear that this would be a team of young men who truly believed that if the fire couldn't "get" them, nobody could.

But before they could start playing again, they needed something pretty small yet extremely fundamental to the game. When Prinz gathered his ragtag group of players around him on that dump of a field, he began the first postfire practice for the Paradise High Bobcats with not a command but a question.

"Hey!" he shouted. "Anybody have a football?"

Back up the hill on the campus, the equipment shed had burned. In each of the player's homes, their personal footballs had burned. This was indeed a football team without a football. Then a couple of the kids had an idea.

"Doesn't Danny have a football?" they asked.

That would be Danny Bettencourt, the team's junior quarterback, who was so diligent that he carried his equipment bag everywhere. It turned out that bag was in the back of the family car when they escaped down the mountain. And it happened to be in the trunk of his car right now. Voilà! A football! Except Danny was in class, so teammates had to run back inside the warehouse and borrow his keys and free the prized pigskin, which Prinz soon held aloft triumphantly.

"Now we can start football practice!" he crowed, before proceeding to direct his kids to a place they would spend the next five months: that field of screams.

Twice a week until the end of May, the Bobcats practiced in those weeds. With the late winter and early spring rains, those piles of dirt quickly became mud. New shoes were ruined in the muck. Borrowed jeans were torn on the brambles. Prinz, whose teams always used the run-centric wing-T offense, couldn't really get traction in this mess, so he wanted to work on his barely existent passing game. But because the field had so many hidden potholes, receivers were cutting left or right and then crumpling to the ground. Overthrown passes often landed in puddles, further damaging the only football they had for the first several practices. So, the coach shut down the passing game entirely, leaving Bettencourt nothing to do besides hand off the ball to his running backs. The field was so bad, one day the entire practice was spent clearing it of rocks.

But at least they were playing. Sitting inside in one of the many makeshift classrooms on this day was a quiet kid who could no longer play. As the school choir practiced from the other side of a partition, senior center Trevor Curtis, the Oakland Raiders fan with a robust frame, raised his voice to be heard. He was part of the graduating class whose season ended prematurely on the day of the fire, and he was still hurting.

"I don't know whether to feel angry or sad. I guess I'm both," he said. "At first, I lashed out, went on a lot of long walks, yelled at people. I was cheated out of the playoffs, and we lost everything in the fire, and none of it seems fair."

Since the fire, Curtis had lived in an aunt's house in the San Francisco Bay Area, then in a hotel room in Sacramento, then in a trailer on a friend's property, and then in a rental house in Chico. The fire had sent his life spinning out of control, with the only possible anchor being that first-round playoff game against Red Bluff. Yet even that was taken from him, and he couldn't tell what hurt worse.

"Right now, right here, let's play that game," he said. "We never had our shot. We never got our chance."

The window on Curtis's high school career, and the careers of the several other Paradise seniors, had closed. Over the years, the school had sent only four players to an NCAA Division I college program, and the few who continued playing the sport usually ended their football days at the junior-college level.

A senior season at Paradise is often a final chance at football glory. This was particularly true for Paradise's most unique senior player.

Her name is Gretchen Cassing.

SHE WAS FINISHING HER Paradise High career as a member of the team's athletic training staff, one of several students who annually assisted head trainer Marvin "Chip" Schuenemeyer, but perhaps more compelling was her second job as one of the Bobcats' kickers.

Throughout the 2018 season, she kicked their extra points. Gretchen was the first female to ever wear a Bobcats uniform. And with an 8-2 record, her team was headed for the playoffs and a chance for Cassing to be one of the unlikeliest heroes in Paradise history.

"The season was a dream," she said. "And then it wasn't."

That dream began the previous summer during the team's summer camp at a two-year school, College of the Siskiyous, located two and a half hours north in the town of Weed. In addition to being an assistant trainer, Cassing was a star soccer player, and during some of the football team's more boring drills, she would retreat to the far end zone and set the ball up on a tee and boot the hell out of it.

"She wasn't just kicking it," remembered Prinz. "She was booming it."

The coach was impressed. He saw more than just a competent kicker. He saw an opportunity for enlightenment.

"I thought to myself, 'What if these boys were exposed to some serious gender equity?'" he recalled. "It would educate them. It would be good for them."

So, one afternoon Prinz summoned her to his lawn chair under a sideline tent. She thought he was going to throw her off the field.

Instead, he said, "Go over to where Coach Wiggins has the team practicing extra points," and pointed to the far goalpost. "Ask him if you can have a shot."

Cassing ran excitedly toward the action. She remained there for all of five seconds.

"You're kidding me, right?" said John Wiggins, whose son Dominic was the team's star running back. "Get out of here!"

Dejected, she ran off, too embarrassed to say anything to Prinz and feeling foolish that she ever thought she genuinely had a chance. Prinz did nothing, knowing that the team would never buy her presence if it was forced upon them.

"That was my first sign that maybe this team wasn't ready for a girl player," she remembered.

Yet after summer camp ended and the team resumed preseason practice at Om Wraith Field, Prinz called her over again.

"I wasn't going to give up so easy," he reflected later. "This was too important."

By then, the coach knew he couldn't mandate that she have a spot on the team. She had to earn it. So he gave her the chance.

"While the rest of the team is down there at one end of the field, you go to the other end and just start kicking extra points," he told her conspiratorially. "The players and other coaches will eventually notice you. They'll see what I see."

They saw, all right. The boys gathered nearly a hundred yards away, and they gawked. They thought it was . . . weird. But they knew that Gretchen was good enough to ring up the extra points that high school kickers often miss, to everyone's frustration. The stage was set. The belief system was put into place.

And sure enough, two games into the season, when she was given uniform number 14, the Bobcats were ready for her. Well, sort of. They initially pushed Cassing to the side during stretching exercises. They would occasionally make snarky comments about her belonging on the cheerleading squad, not the team.

Then she started making extra points, and they shut up.

She kicked the team's extra points for its final eight games, as the Bobcats rolled toward the playoffs. Cassing converted twenty-two of twenty-nine kicks, and that included a couple of blocked attempts. She never cost them a game; in fact, she helped them win several. Gretchen Cassing was bringing the Bobcats into the twenty-first century, and they eventually came to embrace it, including her in the pad thumping and helmet waving, and, yes, even the stretching drills.

Coolest three months of her life. Hardest three months of her life.

"The pressure was awful," she remembered. "The harassment from other teams was terrible."

First, for her, there was the matter of fitting into her uniform so that the five-foot-eight, 140-pound blonde would look . . . stylish?

"Seriously, I'm still a girl, and I was worried about looking gross in the uniform," she explained. "Girls aren't meant to be running around having big old shoulders. I had to be convinced I looked decent so my head would be in the right space. The other girl trainers would braid my hair in the training room, and I would fit it all under that helmet. Then I had to get the uniform looking just right."

Once on the field, Gretchen discovered quickly that no amount of padding could protect her from the opponents' abuse.

"The Paradise fans were wonderful," she said. "The other teams were terrible."

Some would scream gender-based invectives before the extra-point snap. Others would, rather obviously and quite lamely, point at her, then point to the cheerleaders. And after games, in the high-five line, some would put their hands down when she approached, refusing to even acknowledge her as a worthy opponent.

"It was all very intimidating," she recalled. "I had terrible butterflies before every kick. Every time, the worst butterflies. After every touchdown, it took freaking guts to go out there and kick."

But she kicked, and kicked some more, and, after many successful attempts, she would run to the sidelines and jump into the arms of assistant coach Andy Hopper or Jeff Marcus, while the rest of the team howled in delight.

"She was so much fun, we all wanted to score just to watch Gretchen kick," remembered Hopper.

And when she wasn't kicking? That's when Prinz was most impressed.

"One of our players would go down with an injury, and she would run out there to tend to them, in her uniform, still doing her training job at the same time she was doing her football job," he said. "She was the best roster decision I've ever made."

Through it all, Cassing was learning to succeed under fire. On November 8 her education served her well when she had to perform in a real fire.

She woke up later than normal that morning because she was attending a college extension program at Butte Community College in nearby Oroville. Both her parents had left for work. Her little sister had gone to school. She was alone when a friend called and told her that a nearby street was on fire.

"I walked outside my house, and it looked like hell," Cassing recalled. "Literally hell."

A yellow Ford truck drove past. The driver rolled down the window and began shouting at her, "Leave! Leave! Leave!"

She couldn't leave. She had no way to leave. She called her parents, but the calls didn't go through. She ran back inside as flames danced down her block.

Suddenly her father, Tim, pulled up in a truck charred with burn marks and covered in burning twigs.

"Get in, now!" he shouted, and Gretchen, still wearing sweats and slippers, joined him, and they attempted to make their escape.

Except they couldn't go anywhere. Like many in the town, they were stuck in traffic. Fire closed in around them. Panicked cries and propane tank booms filled the air. In the backseat of a Honda stuck in an adjoining lane, a howling woman had gone into labor.

Her father turned to her.

"I don't know if we're going to make it," he said.

Then her father's phone rang. It was her stepfather, Jim Shults, who lived with her mother down in Chico.

"Tell Gretchen to hang in there!" he shouted. "I'm coming to get her!"

You see, Jim Shults had the most valuable of commodities in this deadliest of traffic jams. He had a Harley. His motorcycle weaved through the gridlock and bumped over the burning terrain and ended up next to Tim Cassing's truck.

"Get on!" he shouted. "Let's get out of here!"

Tim told her to go. Gretchen, crying and screaming, jumped on the back of the bike.

"The worst ride of my life," she recalled.

Without a helmet, with nothing between her and her hell, and with her two burning hands clutched tightly around her stepfather's waist, they sped down the mountain through a canopy of flames.

Her ears blistered. Her ankles blistered. She cried the entire trip as Shults continued to rumble around the cars and through the ash. But she held on. Kicking for the Paradise Bobcats had taught her to hold on. Eventually they outraced the terror and found safety in her mother's house in Chico.

Once there, the magnitude of her escape enveloped her, and Gretchen lay on the driveway for two hours looking at the sky.

"In a matter of seconds, I lost my house, my uniform, my helmet, my memories, and I even lost my phone," she recalled. "I lay there in fear that maybe I also lost my father."

Tim Cassing eventually made it down the mountain, and Gretchen was finally able to breathe, or so she thought. She had lost nearly everything, but still she had at least one more football game to play.

"The only thing I could look forward to was that the playoffs were starting the next day, and surely we'd figure out a way to compete in those playoffs," she recalled. "We were playing great, I was kicking well, it was a great memory waiting to happen."

Then the next day, Prinz canceled the playoffs, and Cassing, like many of the team's seniors, was enraged.

"It was awful, we could have won, we would have won, everybody wanted to play, other teams were literally giving us everything we needed—uniforms, their field, everything," she said. "I'm still absolutely mad about it. The decision could have been put off for at least another week. Everybody wanted us to do well. And we never had a chance."

Cassing didn't blame Prinz. She knew her coach had no choice. And, unlike others, her sport was soccer. Still, she wasn't sure she'd ever get over it.

"I'm still mad at the fire," she said. "I'm still mad at everything it took from us."

In return, well, she got a San Francisco 49ers jersey. During the team's trip to Levi's Stadium, 49ers star tight end George Kittle gave her his jersey during the national anthem.

"He was really nice, it was really cool," she said. "But it wasn't my jersey. I'd give anything to have my jersey."

For Gretchen, Trevor Curtis, and other seniors, the fire cut short their last triumphant ride. But on this March day, there were still twenty-two sophomores and juniors out there on that muddy field who had another chance to make up for that lost opportunity. In a cluttered hallway office, there was a feisty woman working to give them that chance.

SINCE THE FIRE, ATHLETIC DIRECTOR Anne Stearns had been working twenty hours a day to rescue the Bobcats' fifteen sports programs. It was Stearns, an energetic eternal optimist, who managed to find a field in Oroville, get donated equipment, and convince Red Bluff to delay last year's playoff game a week so that the football team could keep playing, even though, ultimately, it was not meant to be. It was Stearns who presented Prinz with that option for the sake of his seniors.

"I was a yes, Rick was a no," she said. "But I totally understand why."

Stearns, thirty-seven at the time, ran all sports at Paradise High. After football season was canceled, she managed to salvage the basketball, soccer, and wrestling seasons. Working out of her Chico home and a borrowed office at Chico High, with no facilities or equipment, she lost thirty pounds but pulled off a miracle. Within five days of the fire, the basketball teams were playing on borrowed courts—one of them in somebody's backyard. The wrestling team was practicing in a shed. The soccer team was on a vacant field. She relied on donations and gift cards to scour the sporting goods stores for equipment. It all worked. The seasons were completed. The boys' basketball team even won the league title without a home gym. For her efforts, the former elementary school physical education teacher was named "Sports Person of the Year" by the *Chico Enterprise-Record* newspaper and received offers from other schools. She turned down all of them to remain in the Paradise High muck.

"It is important for me to rebuild what I've worked so hard to keep alive," Stearns explained.

Once she figured out the winter sports, she made it her mission to save the biggest endeavor of all. She had to bring the football team back to life in 2019 for another chance at its lost glory.

"Football is the biggest part of our moving forward with our athletics," she said. "The heart and soul of this town is football, yet all of our fifteen sports came back to their seasons except football. It's really important to get us back up on the hill."

Stearns loved walking around Om Wraith Field during games and marveling at the number of people she didn't recognize.

"Paradise High is so connected to the community, tons of people who come to games have no connection to the school," she said of the weekly crowds that number several thousand. "It's a Friday-night event. Bringing football back to the community is huge for the town's rebuild."

Her first football-related job after the fire was to make sure the most important part of that connection was still intact. She wanted to make sure Rick Prinz was coming back for a twenty-first season.

"Of course, I worried about him," she said, then confided, "I'm surprised all my coaches didn't retire."

Considering that Prinz was paid just $4,000 a year extra to coach the team, she wondered why he would put himself through the pain of a program rebirth. She didn't say anything to him, didn't want to put any ideas into his head; then, one day at the warehouse, Prinz finally confessed.

"You know I was going to retire," he told her with a grin.

"This is the first I've heard of it," she said. "You're giving me heart palpitations."

Later, after Prinz returned to his caged office, Stearns shook her head.

"I cannot imagine rebuilding without him," she said. "He puts his team and community first. None of this works without him."

After making sure the coach was back and the donated equipment had been sorted, Stearns was faced with a more difficult project. She now knew they would be playing, but *who* would they be playing? Because Paradise High had lost almost half of its enrollment, from around a thousand to barely five hundred, the Bobcats could no longer compete in the Eastern Athletic League against similarly sized schools. In fact, they couldn't be placed in *any* league out of fear that they would be playing much larger or smaller schools. So, they were suddenly independents in desperate need of ten games. Stearns sighed, picked up the phone, and started calling.

It was a tough sell. Other athletic directors knew that while the Bobcats might be undermanned, they were always ferociously tough.

"I remember sitting around with several players from our team last year and asking them the name of their biggest rival," she recalled.

"Going around the room, they each mentioned a different team: Chico, Pleasant Valley, Enterprise"—a high school in Redding, California—"until I realized that they had eventually mentioned every team on our schedule. It was then that I realized, this was us against the world."

Her second obstacle was to find a team that was unafraid of the potential controversy that any outcome against the depleted Bobcats would cause.

"If you lose to poor Paradise, it doesn't look good for you," she said. "But if you beat poor Paradise, then you've beaten a team whose town burned down, and it *still* doesn't look good for you. Bottom line, you play us, you're in a no-win situation."

Which was why every school that Stearns called turned her down. She focused on smaller schools, given that Paradise was going to lose a lot of its enrollment, but they wanted no part of the Bobcats. She tried bigger schools, but those schools all wanted to play Paradise on their home fields. And *that* posed a problem because she didn't know how the team would get there. "All our bus drivers moved out of town after the fire," Stearns explained. She remembered thinking, "If we are going to make these games, I'm going to have to get my bus license."

She begged. She bullied. She brought herself to tears in recounting her story of watching the tires melt on a school van before making her escape down the hill.

"I used the same sad story again and again: 'Our town burned down. Surely you can help us rebuild by coming to our field for a football game?'"

She pleaded, other athletic directors listened, and finally it was settled. Almost. Stearns was able to schedule four home games and five road games. She needed one more, but it was the most important game of all. She needed someone to come to Om Wraith Field on August 23, 2019, for what would arguably be the most emotional high school game in the country this season: Paradise's first step up from the ashes. Talk about a tough date. Talk about a no-win situation. You'd almost to have bribe a team to get it to show up.

That's almost what happened. BSN Sports apparel company offered to donate $5,000 in Under Armour equipment to Williams High, a school of about 558 students located in a town of 5,324 about seventy miles south. In exchange, the Williams Yellowjackets would agree to walk into the tumultuous first steps of the rebirth. School officials there later rejected the monetary offer but took the game.

"We'll take it any way we can get it," said Stearns.

After several months on the rocky pasture, on May 20 the team finally moved to a better location. It still wasn't a football field, but at least it wasn't littered with potholes and soda cans. It was Marsh Junior High, in Chico, about fifteen minutes from the warehouse and a million miles from regulation.

The first thing you noticed about the field was that the parking lot served as the locker room. The players, who by now had acquired donated practice gear, got dressed outside their trucks. The second thing you noticed were the bathrooms. There was one Porta-Potty on the corner of the parking lot. The one thing you didn't notice was the water. That's because there wasn't any. No hoses, no drinking fountains. The team brought in its own water from the warehouse.

Squeeze through an opening in a fence and finally see the field and, well, that's all it was. A field. One giant patch of grass with no lines, no yard markers, no end zone, no borders. A training table had been set up on the edge of the field. A couple of kids hopped up to let two young women tape them before practice started, whenever that would be. Practice never began on time because Prinz never gave a time—since he didn't know how long it would take the kids to drive over in traffic. It sometimes took the coaches two trips in their trucks to collect and deliver everybody. Maybe more. Anything to keep the kids from giving up and quitting on the spot because they couldn't get a ride.

But when practice started, it started. The players finally and joyfully stepped onto a field unburdened by potholes and rocks. They ran wildly around the unmarked patch, gaining invisible yards while pushing toward nonexistent end zones. They hugged for no reason other than the fact that they were together. They shouted the unintelligible cries of the relieved.

"I'm not an administrator, not a counselor, not a psychologist," said Prinz, eyeing the initial chaos with a shrug. "I'm just a football coach who decided to stay and do what I could to help our town rebuild. If this is what it takes, then so be it."

They did jumping jacks like they were jumping up to touch the stars. They hammered blocking bags like they were trying to destroy them.

Standing nearby was Andy Hopper, in his twenty-fifth year on the coaching staff, a giant of a man who on this day was near tears. He'd lost everything in the fire and didn't have insurance. Now he was living in a trailer. He worked an hour away in a juvenile detention center. This was only spring football on an unmarked field. But he was still here.

"I'm still lost. I don't know where the Lord wants me to go, but I do know one thing," Hopper said. "This is where I need to be. This is my breath of life."

Stetson Morgan took a handoff and ran far past the collection of defenders and into that imaginary end zone.

"I get out here, and everything bad goes down, and everything good comes up," he said.

Taylor Brady, a sophomore receiver, ran downfield as if being chased. His family lost everything in the fire, so he now lived ninety minutes away in a house he shared with five others. He awoke at four thirty to get to the bus stop in time to make it to school a couple of hours later. With football practice, he wouldn't return home until long after dark, but he didn't care.

"My life gets exhausting," Brady said. "This makes me feel normal."

Andy Hopper's son Riley, a six-foot-two offensive lineman, stomped around the middle of the line as if he were blocking with full pads.

"I lined up and looked to my left and saw my buddy Ray, I looked to my right and saw my buddy Nate, and I finally felt at home," said Riley, who was still just a freshman. "It made me forget, and it reminded me all at the same time."

Back at the warehouse, a school official watched the Paradise kids begin the journey back and smiled.

"Paradise football is the tip of the spear," said principal Loren Lighthall, who later took a job in another town because his house had burned down, and he and his wife were weary of trying to raise seven children in a 1,100-square-foot temporary apartment. "It's the reason kids come to school. It's the reason the town comes to the games. It's saving this place. We get depressed, we can't talk our way out of it, so we turn to football."

Here on a late-May afternoon, playing football on this non-football field, a couple of weeks before they would be allowed to return to their home grass, and three months before that monumental first game, the Bobcats acted like they were already in the Super Bowl.

They huddled and shouted, "Brothers to the bone!"

They huddled again and shouted, "Pick your brother up!"

The heavy lifting began.

FIVE

The Coach

JANUARY 14, 1959

But first whom shall we send in search of this new world?

The coach was born in a car.

The life of Rick Prinz has been one of makeshift survival from the beginning, from the cold moment on January 14, 1959, when his mother went into labor in their home in the tiny central California mountain town of Pinehurst. His father, Bill, helped Carole into their Pontiac station wagon and barreled down what was then Highway 69 toward the nearest hospital, an hour away in Exeter. But Rick wouldn't wait. She began screaming from the backseat that she was about to give birth. So, halfway down the hill, Bill pulled into a bar and retrieved Rick's grandfather Ted Clem, who jumped behind the wheel and drove while Bill hopped in back and delivered Rick. When they finally reached the hospital, the two men hustled the new baby into the emergency room— inadvertently leaving Carole in the car. Whoops.

"Tough start," deadpanned Prinz.

Sixty years later, on his football team, Prinz utilized a fast-moving offensive scheme that honors his roots. It's called "Speedwagon."

"If only they knew," he said with a laugh.

One of five children, Prinz comes from tough stock. His father was the son of farmers, while Carole's parents owned restaurants and a grocery store. When Bill was fifteen, he joined the US Forest Service, then, at eighteen, enlisted to fight in Korea. He was wounded three times on the front line, taking shrapnel in his back and his cheek. Then he was captured by the Communist Chinese during the crucial troop withdrawal in the Battle of Chosin Reservoir. He was being marched into China when he and a couple of friend broke ranks and scampered into the night. They knew they had nothing to lose, for the alternative was death.

The mad dash worked. Bill escaped and eventually returned to his beloved Sierra Nevada mountains to rejoin the US Forest Service as a firefighter.

"You're eighteen years old; how do you just decide to run for your life?" Rick once wondered aloud to his father.

"When you see your buddies march off into the forest and never come back, you get real brave," Bill responded.

Rick inherited not only his father's resilience but also his stoicism. Bill never talked in detail about the war or about being a POW. He never divulged much about the escape. Rick eventually stopped asking and learned a lesson that he carries with him today; one that would help him quietly lead his football team out of the unimaginable flames.

"It's not what you say, it's what you do," Prinz said. "Don't talk about it, *be* about it."

Rick spent the first six years of his life in a small cabin in Pinehurst. He was an outdoorsy kid who couldn't stay out of trouble. Considering he lived in a US Forest Service village, he gravitated toward the worst kind of trouble. Once, when he was five, he was told specifically not to play with matches. So, what did he do? He marched into his family's laundry room and lit one match, then another, then another, then *whoosh!* His father's Forest Service shirts caught fire. Young Rick ran outside and turned on a sprinkler and tried to douse the flames. It didn't work. The only thing that kept the house from burning down was their good fortune to live next door to the neighborhood fireman.

Then, Rick knew, he had to face his father. He waited on a lawn chair in the backyard for Dad to show up. The boy was trembling and sweating so hard that Bill took one look at him and did nothing. He knew Rick had already punished himself. Many years later, this is how Rick disciplines his players after a fumble or interception or missed block. As they come off the field in terror, he'll just stare at them and shake his head. He doesn't yell at mistakes. He doesn't scream at blunders. He knows his players are feeling far worse than he could ever make them feel. If they're not hustling or screwing around on the sidelines, that's a different story. That's push-ups or sprints or extra work. He doesn't completely spare the rod, because his father never completely spared the rod when he thought his son deserved it.

"He didn't use a rod, he used a razor strap," Prinz recalled. "And just hearing him get the strap was as bad as feeling the strap itself."

When Rick was six, the Prinz family moved to the small central California farm town of Willows. He was a rowdy kid in need of an outlet. He craved attention. He was in trouble a lot.

His father would hit him with the strap, or rap him on the head or neck, partially to get his attention if he thought he didn't seem remorseful, and partially to toughen him up. When the transgression was serious, as when Rick accidentally killed the family hamsters by forgetting to give them water, the punishment was more severe. For that, he had to sit on a dirt pile in the sun for four hours while pondering the loss of life.

"My dad had faced death many times; there was no backing down in him," Prinz recalled.

There was the time his father—all five foot ten, 180 pounds of him—chased a mechanic around the car with a wrench because he thought he was being ripped off. He brought the same unbridled energy to the family home. Often Bill Prinz would work all day for the Forest Service and then come home to work on the house, putting on siding in the dark. Rick's brother held up a light while Rick swatted mosquitos off his father's back. His mother was tough, too, though in a different way: she was a high school principal's secretary who insisted on making breakfast every morning for her five children.

Rick's parents instilled in him a work ethic that survived many scrapes, as well as a humility that endured many spills.

There was the time his father was building a driveway in front of house and ordered ten-year-old Rick to retrieve a heavy chain and bring it to him in a wheelbarrow. Rick decided to be a tough guy and carry the chain himself. He fell hard, the chain landed on top of him, and he'll never forget the only two words father spoke as he lay there immobilized:

"Get up."

Then there was the time his father was roofing the house, and Rick was asked to carry one sheet of tin at a time up the ladder because of the presence of an exposed electrical wire that couldn't touch the tin. Rick decided to show off for his then-girlfriend Veronica, brought up two sheets, and the wind blew one of them into the wire. Rick was literally shocked off the ladder and toppled to the lawn. His father just looked down at him as he lay there in terror and repeated those same two unforgettable, unsinkable words.

"Get up."

Prinz marvels at how many times he has uttered those same words to a player who has tried to show off and taken a tumble. He is amazed at both the power and forgiveness contained in those two syllables. During low moments in the wake of the fire, he even repeated them to himself.

"My father was the strongest man I've ever known," he said with the utmost respect.

One of Rick's first pets was a wolf. His father brought it home, and Rick learned to play with it. Bill actually had more trouble keeping a leash on his troublemaking son. Once, in a third-floor Willows classroom, a frustrated teacher walked in to find Prinz standing on a desk.

"Daryl Adams told me to get up here," Prinz claimed, pointing to a buddy.

"Would you jump out of the window if Daryl told you to?" the teacher asked.

"Will I get to go home if I do?"

"Sure," said the teacher.

Prinz looked at Adams and grinned.

"Jump out the window," Adams said.

Prinz promptly flung himself out of the window, landing in a heap on the grass below. Of course, he wasn't hurt. Of course, he walked home. And, of course, he thought of his father.

"At least my dad wasn't there to tell me to get up," he said.

Another time, Prinz and friends dashed out of the high school building in the middle of the day and locked everyone inside. School administrators didn't find it as funny as Bill and his pals did. Nor did the normally stoic Bill Prinz, who fetched the strap.

Throughout his turbulent childhood, the one constant for Prinz was sports. He could always be completely himself, without remorse or retribution, when he was on the field.

He remembers being a first grader in Willows, the new kid on the block, standing in the bus line and crying because of his shoes. The other kids wore Converse sneakers; Rick wore dress shoes. Every day at recess, when the other children ran around and played, he would be left behind because of his shoes. They kept falling off, and he would have to chase kids in his socks. He begged his parents for two months for a pair of Converse, which cost $10. They finally found the money and bought them for him.

"And from that day on, man, I could fly," said Prinz.

When he was in second grade, Rick realized his flight path would be football. He was given a football as a Christmas present, and he used to lay in his bed for hours, tossing it up and catching it. Eventually an older cousin took him to the backyard and showed him how to run some receiver routes. Then at school, they played a passing game called splatterball, which entailed passing a football down the field without stopping or huddling, and Prinz was hooked.

"I wanted to be a football player," he remembered. "That was all I wanted to be."

Willows didn't offer youth football, so he didn't start playing until seventh grade, when his parents drove him to nearby Orland. He joined the Willows team and four years later graduated as an all-league linebacker and the team MVP—a five-foot-ten force of nature even when nature fought back.

Rick spent much of his free time working for local farmers in irrigation and construction. He once tore down an old church, hauled its pieces over to a salvage yard, then was nearly crushed when the frame collapsed on him. He sliced up his arm in one auto accident and was lucky to climb out of a ditch after another.

"I was seventeen years old, and I didn't really know what I was going to do with my life," he reflected. "I just knew I was going to Shasta College to play football."

Shasta College was a community college in Redding, about eighty miles from Willows. But Prinz never made it that far. He was sidetracked by Veronica Taylor. She was from a large, close-knit Willows family, and Butte Community College was only forty minutes away. And that was that—at least according to the future Mrs. Prinz.

"I told Veronica I'm never going to Butte College," Rick recalled. "Veronica said, 'You'll go to Butte College.'"

They'd met in first grade, went steady in seventh grade after Prinz threw a plastic ring at her, but broke up a week later in a message passed to him by her friend. They were in a movie theater. Veronica's friend handed back the ring and told Prinz it was over. But it was just beginning. Five years later, when they were both seniors, they found themselves together at a church Halloween dance. Veronica pulled Rick's shoes out of a pile and claimed them, and him. Five years later, they were married. Thirty-nine years and four children later—all boys—she still walked the sidelines with him during games and kept the offensive stats.

"Romantic, huh?" he said.

As preordained, Prinz went to Butte College, where he was immediately installed as a starting linebacker. That lasted two games. In fact, his entire college career lasted two games. In that second game, he was rushing in to block a punt, an offensive lineman hit him low, his left knee ripped apart, and he knew it was over. Actually, he'd known it was over before it was over. For one of the first times in his life, this level-headed young man who had no use for dreams had a prescient dream.

"The night before the game, I had a vision. I thought, 'What am I going do when I hurt my knee?'" he said. "It all came true."

He underwent reconstructive surgery the following Monday, and, despite being in a cast up to his hip, insisted on returning to Butte a few days later. Big mistake. Prinz started sweating so badly that he jumped into the campus swimming pool. He dried off, crutched home, and was immediately admitted to the hospital with a staph infection. He remained there for twenty-six days, lost fifty pounds, and was stuck in the cast for six weeks. It became so uncomfortable that at one point, his father ripped off the bottom of the cast to give him relief. But there could be no relief.

"This was the low point of my life," he recalled.

FOOTBALL WAS DONE. AND SOON college was done, too. Rick dropped out and went to work at a local fiberglass plant for the next eighteen months. He packed it. Carried it. Put it in a bag. Sealed it. Sweated through it. And when he was finished, he drank and partied with his old high school friends to ease the physical and emotional pain. One particular night he was drunk and staggering when, for whatever reason, he suddenly remembered the words of a former Willows coach named Bob Moore. The coach had cornered the cocky senior after school one day with a question.

"If this building fell on your head right now, where would you go?" he asked him.

"I don't know," Prinz answered.

"*I'd* go to heaven," said Moore.

"You're crazy," said the teenager. "How do you know that?"

"You need to figure that out," said Moore.

A year later, that question still touched Prinz, as did the quiet grace of another mentor, Willows High School's junior-varsity coach, Terry Franson. Soon after Prinz's knee injury, Franson called him and asked if he wanted help with his rehabilitation. For weeks, every morning at six o'clock, they would meet at a nearby field, say a prayer, and jog around the turf. Soon the prayers felt as important as the running. Prinz believed he was exercising his soul.

After one night of drinking, Rick looked around at his world of debauchery and excess, thought of these two men, and made a decision.

"I said, 'There's got to be more to life than this,'" he remembered.

He started attending the First Baptist Church of Willows. No matter how hard he partied on Saturday night, he was in a back pew come Sunday morning. One afternoon after services, his close friend Jeff Fleming burst into his house and said, "You should meet this new youth pastor!" Prinz responded, "What's a youth pastor?"

Soon he was attending youth group classes led by Lee Talley, a cool dude with a chill message. He taught tough-guy Prinz how to be a Christian without being "religious." He could be rowdy and funny while still being virtuous. He could be himself while still being with God.

One night Veronica, who grew up in a Catholic household, played him a recording from a Christian comedian who showed, again, how someone could be both cool yet Christ-like. That did it. Prinz was sold.

"That night, I was laying there, and I started to pray," he said. "'I want to be like Coach Franson and Coach Moore. I want to be saved.'"

So, at age twenty, with a bad knee, a job at a fiberglass plant, and no discernable career, Rick Prinz became a Christian. After which, of course, he threw a party for twenty-five people, only this one included Moore and Franson and testimonials. The two coaches have maintained a special place in Prinz's life. Prinz calls Moore before every game to pray. Franson, who became a legendary track-and-field coach at Azusa Pacific University, made several visits to Paradise to stand alongside Prinz during the comeback season of 2019.

"Becoming Christian was the beginning of the rest of my life," said Prinz.

He finally quit his job and enrolled in Fresno Pacific University, a Christian university that Talley attended, and earned a degree in Christian education. He was going to be a youth pastor just like his friend Talley. He married Veronica in 1981 at his parents' house and, later that year, began another long-term relationship: coaching football.

After two years as a varsity volunteer assistant at Willows High, Prinz made his head coaching debut as the leader of the school's 1981 freshman team while still finishing college with a correspondence course at FPU. Yes, his mind was still set on becoming a youth pastor, but his heart was always in football. Nobody, not even Rick, had any idea how much the heart would win out.

There were only twenty kids on his first team, and their star running back was called up to the junior varsity midseason. Though shorthanded, his Willows High Honkers (named after a flying Canadian goose) went undefeated at 10-0. The twenty-one-year-old Prinz led with his body, playing quarterback during practice. He led with his heart, once being assessed a fifteen-yard penalty because he'd raced into the end zone to celebrate a touchdown along with his young squad.

"That season hooked me," he recalled. "I wanted to be a coach. It was on."

Yet Prinz and Veronica had their first of four children, Billy, in 1985. Prinz needed to support his family, and youth pastor seemed a surer bet. Prinz earned his master's degree in Christian education at Biola University, a private evangelical school in the suburban Orange County city of La Mirada, before returning north to begin his ministerial career at the Chico Neighborhood Church. At the same time, he kept one toe on the football field by coaching at Chico Junior High, working with the ninth-grade team as an assistant coach for running backs and linebackers.

In 1989, looking for a church change, and a better football job, Prinz began a stint as youth pastor at Magalia Community Church, located just outside of Paradise. A year later, he started what was to be a two-year hitch as a volunteer assistant at Paradise High but wound up staying there for the next thirty-two years.

"Once I got here, I always felt like this was where God wanted me to be," he said. Small kids. Difficult schedule. Unruly conditions. All those combined to make Paradise Prinz's kind of place. Plus, it was a challenge: the once-proud program had fallen into disarray.

The Bobcats began playing a full schedule in 1956, and, by 1964, they were league cochampions. Over the next seventeen seasons, Paradise High posted eleven winning records, including a sectional championship in 1980. In Prinz's first season as a volunteer assistant, they were 8-2. Then the hard times struck. Over the next eight years, during which Prinz served as everything from freshman coach to varsity assistant, the program didn't have one winning season, going a combined 15-65 during that time. Nevertheless, Prinz impressed constantly with his unique mix of aggressiveness and composure—think rowdy youth pastor—but he never had the title to give him the ultimate influence. He was so immersed in football that he finally quit his job as youth pastor to work at the school full-time as a coach and health and technology teacher—"I figured I could minister these kids better on the football field"—but he still couldn't get a crack at the head coaching job.

Finally, coming off a 2-8 season in 1998, coach John Luciano stepped aside, and school officials decided it was time. Principal Jeff Dixon approached Prinz in a school hallway with a request.

"I want to hire you as the football coach, and I want you to build a program," he said.

Bobcats culture was born.

"I was fired up, determined to turn the program around, not embarrass myself," Prinz recalled. "First rule, we don't hope to win, we *expect* to win."

He brought his work ethic in the form of spring training, summer programs, three-hour practices, and even, initially, illegal workouts to switch from Luciano's modern pro-style offense to Prinz's traditional running wing-T offense. Where Luciano's regime was also pro-style in its detached approach, Prinz brought a personal touch by talking to the kids, getting to know them personally, joking while he was demanding, connecting while he was leading.

"Yeah, you want to teach discipline, but you want to do it as a labor of love," he explained.

"We Just Hit People" was born.

The Paradise motto, now printed on everything from caps to T-shirts, was devised and instituted by a coach trying to change an attitude.

"Right away, I felt like we weren't being physical enough," Prinz recalled. "Willows tore people up. Paradise *got* torn up. Every coach talking about being a physical team; well, we were going to be a physical team and not just talk about it. But we needed a slogan, something for the kids to own. 'We Just Hit People' is more than just words, it's a mind-set."

"CMF" was also born.

His first season, a kid showed up with a "BMF" sticker on his helmet. The kid told Prinz it meant "Bad Mother Fucker." Prinz hated the words but loved the sentiment.

"I'm looking for ways to say to kids, 'We've got to be tough, aggressive, we've got to play like we're crazy, out of our minds,' but how am I going to do that in a way they can easily understand it?" Prinz said.

"BMF" gave him an idea. How about "CMF"? he proposed. Only, he wanted it to stand for Crazy Mountain Folk. He tried out the motto, and the kids immediately translated it to "Crazy Mother Fucker." He kept insisting it remain clean. The kids finally listened. Their parents listened. The town listened. Today "Crazy Mountain Folk" is how most interpret CMF, which has become everything from a rallying cry, to inspiration for a tattoo, to the name of a town business.

"If used in the proper way, it defines our mentality," said Prinz. "When we're at our best, we play like crazy mountain folk."

In Prinz's debut season of 1999, they were just plain crazy. Changing the culture was hard. The Bobcats played out of control. "Wild mountain boys," they were called. Prinz wanted them to be tough, but they felt that gave them license to be nuts—and that wasn't good. As the *Paradise Post* described the end of one loss to the Wheatland Union High School Pirates: "In the final five minutes, the Bobcats committed three personal fouls. One for swearing, one for a late hit, and the other was for choking an opposing player."

The Bobcats went 4-7 in Prinz's first season but somehow refrained from choking opponents long enough to actually make the playoffs for the first time in a decade. The next season they went 9-4 and advanced to the sectional final for the first time in twenty years. In 2002 they were undefeated section champs, 12-0, the first of Prinz's six sectional titles. Equally as impressive has been a Prinz-inspired junior Bobcats program in which the town's youngsters learn to run Prinz's offense shortly after they learn to throw and catch a football. Long before the fire, Prinz had staked his claim as the most influential figure in Paradise.

"Rick Prinz *is* Paradise football," said athletic director Anne Stearns.

But success has come with a physical price. Prinz suffers from atrial fibrillation, an irregular, often rapid heart rate that can increase the risk of a stroke. He'll wake up in the middle of the night before a big game and feel like his chest is exploding.

"I don't feel anything," he explained, "but my heart—boom, boom, boom, stop, fast beat, pause, fast beat, pause, going crazy, going crazy. I'm like, 'This has to stop.'"

If it lasts more than hour, he has to go to the hospital out of fear he may suffer a stroke. He's made that visit four times. He's surprised it's not more.

"Stress can bring it on," he said. "Lucky me."

With his usual impassive manner on display even amid danger, Rick Prinz is forever his father's son. He wishes his father could have lived long enough to see his success. He wishes his father could have seen how his toughness rubbed off. He wishes he could give his father a hug. He wonders if his father would hug him back.

Prinz lived twenty-nine years without ever hearing his father say, "I love you." Then, at age fifty-nine, Bill suffered a heart attack. When he saw his children afterward, tears streamed down his sagging cheeks.

"Go on, hug him, he's crying," said Rick's mother.

"I don't know what to do," whispered Rick.

But he leaned down and hugged him anyway, and his father stiffened. Then Rick gave the pep talk of his life.

"I want to tell you something," he said. "There's a lot of other guys I've looked up to, but I looked up to you the most, and I love you."

The weakened Bill Prinz struggled to his feet and said, "Thank you. I love you too."

Ten years later, on March 31, 2000, Prinz was approached in a school hallway by a campus security officer telling him he needed to go to nearby Feather River Hospital. The officer didn't say why.

When Prinz arrived, he saw paramedics administering CPR in a nearby ambulance. He looked through the window. It was his father. Earlier that morning, while standing over a sink in the family kitchen, Bill's carotid artery ruptured, filling the basin with blood and dropping Bill to the floor. He never had a chance. He was removed from the ambulance on a gurney. Rick followed him into the hospital emergency room, standing there while paramedics and then doctors continued to work on him. He gently put one of his father's distended fingers inside the railing, held his hand, and then he felt another presence.

"I swear there was a third person in the room," Prinz recalled. "I swear someone was standing right here. I think it was the Lord. Because I looked down, and my father was gone. The best man I've ever met."

Rick Prinz wanted to stay by his father's side forever. He wanted to grieve all night. He wanted to lose himself in the emotion of saying farewell to his hero. Instead, he left the hospital and drove back to his parents' house. The neighbors would be visiting soon, and someone had to clean up the blood.

SIX

The Summer

SUMMER 2019

From the lowest deep will once more
lift us up, in spite of fate.

Lukas Hartley took a handoff and scampered up the field to . . . where? Where did he go? He literally disappeared into the humid darkness.

Danny Bettencourt dropped back and fired a pass to . . . who? Who caught the ball? Did *anybody* catch the ball? Where did they go?

It was early August, and the Paradise Bobcats were so thrilled to finally be playing in Om Wraith Stadium for the first time in nine months, they were thrilled to play in the dark. It was the second night of the three-day summer football camp, and they had been practicing since nine in the morning. Now it was nearly twelve hours later, and they still weren't ready to quit. They were in T-shirts and shorts, but still they hit. Though exhausted, still they sprinted. And, while the giant field lights remained shut off and darkness descended, still they played.

"One guy has the keys to the lights," explained Prinz. "But if he doesn't want to take the time to turn them on, we'll keep hitting in the dark."

That guy was longtime defensive coordinator Paul Orlando, and while, yes, he possessed the keys to the lights, it wasn't because he was a coach, but because he was also the school's groundskeeper. The fire destroyed not only Orlando's home but also his landscaping business. Orlando needed a job, and the school district needed someone to maintain the high school building and grounds, so here he was, showing up before dawn to do his school yard work and staying after dark to coach the football team. And as unorthodox as it sounds, yeah, he liked the team having to feel its way through the plays.

"It's dark?" he said jokingly through the blackness.

The players forged ahead with a passion not seen at their practices at Chico Airport and Marsh Junior High. It was a passion reflected in the roadside banners they passed on the drive up Skyway Road to the field. The flapping signs were little promises nestled amid the rusted and twisted debris that still blighted much of the decimated town.

"We're in This Together."

"Don't Stop Believing."

"We Are Ridge Strong."

Slowly, the players were once again embracing the quirkiness of Bobcat Strong, so much that they even resumed a tradition of singing to their aging coach whenever he made a blatant mistake in calling the plays or called someone by his wrong name. They stopped the play and, in unison, sang , to the tune of "Old Man River," three taunting words: "Old man Prinzer!!!"

"Welcome to Paradise," quipped assistant Andy Hopper.

"We're just flying by the seat of our pants," said Prinz.

The players didn't just tolerate the dark; they thrived on it. Many of them had been in the Bobcats program since they were children, running the same wing-T offense that Prinz instituted with the high school team. They knew these plays like the grass-stained backs of their hands. The darkness was no big deal.

"You have to understand, we've all run these plays since we were ten years old," said Hartley. "We could do it with our eyes closed. Nobody notices that it's dark. Even though we're not in our houses that burned up, we all feel like we're home."

Hartley, a stout six-foot-one, 195-pound kid with a big smile, had lived with older brother Logan in a rental home near the base of Paradise. He grew up in Paradise. He loved the closeness of the pines, the tightness of the community, the culture of the ridge. He loved it so much that on the morning of the fire, he was up at three o'clock, sitting on his back porch, waiting for an early sunrise.

"Except on this day, the sun kept taking longer and longer to come up," he recalled. "Then at one point, I looked down and saw a big chunk of ash on my black car. There was no sun, but the sky was burning."

Even then, Hartley wasn't worried. Like most Paradise residents, he had lived his entire life with the threat of fires. Paradise folks shrugged at the thought of fires in the same way that Southern Californians brushed aside concerns about earthquakes. You knew they were coming— someday—but you'd survived them before, and you'd do so again. Hartley shook his brother awake and told him there was fire outside, but it was no big deal. For proof, he pulled out a photo book.

"I've got a hundred photos of helicopters roping water on us; lots of fires in the past. It wasn't serious. It's never really serious," he assured a worried Logan.

To placate his brother, Lukas agreed that they should wait in their separate cars until the danger passed. So they sat in their driveway, watching.

Then they saw the blaze engulf the trees surrounding their backyard.

"We saw the flames on the treetops," Lukas recounted, "and I thought this was it, this was really happening, the house was going to go up for sure. We ran back inside, loaded up the dogs, a laundry basket with some clothes, one or two pictures. We saw the house catch on fire, and we looked at each other, like, 'I guess it's time to go.'"

The fiery trip down Skyway was the easiest part of his journey. After staying with friends for a couple of weeks, with nowhere to live, Hartley left California to join his father in Wilsonville, Oregon. He hated it immediately. The school was bigger than Paradise High, the stares were frequent, and he felt lost. He hated it so much that he blew off school. And every weekend, he drove eight and a half hours each way to return to Paradise to be with his friends.

One day Lukas received a text from his brother. It was a photo of his Paradise number 20 jersey. That was the number he wore in honor of his half brother, Joshua, who'd once worn that number but had died three summers earlier in a drowning accident at age seventeen. In fact, that number was the reason that Lukas, who used to be an offensive lineman, became a running back. Coaches wouldn't let a lineman wear a number that low, so he asked to switch positions—and thrived once he did.

"That number is my life, and once I saw it again, I knew I had to wear it again," he said. "That photo told me I had to come back to Paradise."

Thus, after spending the spring semester unhappy in Wilsonville, he and his brother found an apartment in Chico and convinced their father to let them move back. On the last day of school, in May 2019, Lukas Hartley returned home.

"Everybody in Oregon said, 'There's nothing back there. Why are you going home? Why would you do it?'" he recalled. "I told them, 'You can never put my shoes on. If you would, they would fit you like my town fits me.'"

In the darkness on this summer camp night, a breathless Hartley pointed to the shadowy figures of his teammates as they slapped him on the shoulders and back. He shook his head.

"There's no place like Paradise; it's the secret nobody else knows," he said. "It's us. It's my friends. It's the football team. It's a bunch of kids who might not even come to school if it wasn't for the team. It's such a close connection."

This being the end of the second and final summer camp practice, Hartley, one of the team captains, was charged with holding a players-only meeting on the now pitch-dark field.

"We're going to leave you guys on the field to get together and do the right thing," instructed Prinz.

Hartley gathered the thirty-nine varsity players together and walked them down to the sideline across from the brick home stands. They were standing in front of a fence adorned with advertising banners trumpeting Paradise businesses that had been closed, such as Darlene's Fine Chocolates, Northstate Carpet Cleaning, Paradise Bikes. All of them, gone.

They were standing in the shadow of a scoreboard that was partially melted. They were standing across from a sectional championship plaque that was totally burned. Their young voices echoed off the trees. They were talking about sticking together through the upcoming storm.

"We're all equals here!" Hartley bellowed. "I don't care how fast you run, I don't care how much time you spend on the field! You wear the same color jersey, you are family, and together we will fight!"

Then they literally started fighting, breaking into individual wrestling matches, whooping and hollering and bringing Prinz down into the friendly chaos to break them up.

"C'mon guys!" he shouted. "I leave you alone for two minutes . . ."

Later, when they were out of earshot, Prinz smiled and whispered, "They've been pulled apart, they're finally back together, they're reacting to each other, they're ready to hit each other. We'll be in pads in two days; they'll get their chance."

PRINZ THEN ORDERED EVERYONE UP the hill and herded them into the Paradise gym. They passed a row of smoke-damaged training tables and other blackened equipment. They entered the gym where, for the second straight night, they would eventually lay down on the floor in what was essentially a football sleepover. It was hot and awkward and uncomfortable. Yet for many, it was the two best nights in many weeks. These players had been together since huddling up as children in the Paradise junior football program. Long before the fire, they had formed their own little football community. They were one another's best friends, closest allies, consistent confidants. Like the town itself, their Bobcats community had also gone up in flames.

"It's the first time most of us have spent a night in our hometown in nine months," said Hartley. "We're not in our houses that burned down, but we're with each other, and it feels like home."

It's your typical small-town high school gym, so tiny you can smell the sour sweat and feel the hardwood dust. There are several rows of bleachers on one side, a stage on the other side. Behind the stage is a large physical education locker room with two adjoining side rooms. One is a musty walk-in closet with lockers looming above benches covered in dirt-caked backpacks and dirty T-shirts. This was the official football locker room—and hence the reason why the Bobcats dressed in the parking lot. Just not enough space in here.

The other room was Prinz's office, which he shared with phys ed instructor Seth Roberts. The coach had an old metal desk that he wouldn't lock because once it wouldn't respond to his rusted key, and he had to physically wrench the door open. He sat in a worn chair that screeched across the linoleum floor. Aging and wrinkled football photos decorated the wall, along with a framed Coach of the Year certificate. There was a tiny refrigerator crammed with bottled waters, and an adjacent bathroom that looked like it hadn't been cleaned in years.

"This might not look like much, but none of it burned down," said Prinz, gesturing to the three rooms. "Think about that. That's got to be a sign."

Hauntingly, every backpack or pair of shorts or muddied cleats in these cluttered rooms hadn't been so much as touched for nine months. Nor had the players set foot in here since the fire. Prinz told them to come and reclaim their possessions, but most just left them here, as if the memories of a former life were too difficult to embrace. It might indeed have been a sign, but of what?

Prinz's office also contained a cot. He would sleep here, down the hall from the players, who were to be monitored by several coaches in their own cots spread out around the gym. This camp used to be held in the nearby mountains at a junior college where the players got to sleep in the dorms, but this year there wasn't enough time or money to make the arrangements. In fact, this year the usual $200 camp fee was waived. The players were asked only to bring any sleeping gear that they owned. For most of them, that meant brand-new sleeping bags or borrowed air mattresses—even a used tent erected at midcourt. Before the weary players settled in for the night, one by one they stopped by the stage to share their stories with a reporter. They knew this all looked uncomfortable and strange. They wanted the world to know it was actually the wonderful kind of normal.

Josh Alvies, a thick senior linebacker with short hair and a sturdy smile, and one of three players whose houses survived the fire, marveled that he finally had company.

"I'm living in a ghost town," he said. "This team, our brotherhood, this is what is going to bring our town back."

A cockroach crawled up his leg. He didn't even notice. He was feeling a different sort of energy.

"Coming back to the field gives me chills every time I walk down through those stands," Alvies said. "We're all about those chills."

Tyler Harrison, nicknamed "Junior," a slender, curly-haired running back, had just returned from living with his parents in San Diego for seven months. The fire destroyed his family's house, and he lost everything but his football gear. But he returned anyway, to live with his disabled grandmother in a Chico apartment. His story mirrored the other tales of those who were displaced and returned: that of a close-knit rural community where many who lived there found a niche, where kids of all sorts bonded in their rough-hewn isolation, where the rural trappings seemed far more beautiful than the country's most gorgeous cities. Just ask Harrison. Even San Diego wasn't Paradise.

"It's hard to explain. My parents aren't here—I miss them every day—but Paradise is still my home," he said. "I didn't enjoy San Diego. I didn't fit in. It was bad, man. I was freaking out not being here."

The six-foot-two sophomore admitted that he also sometimes freaked out from not living with his parents. They FaceTimed every day, but he didn't know how long they could be apart. He didn't know how long his season would actually last. But he knew he had to give it a shot.

"Everything is bad, man," he said. "But I need to be with my brothers."

Ashton Wagner, a giant sophomore defensive lineman and occasional running back, talked about his initial move to Fresno. That lasted about a month and a half. He came back with his mother, settling into a home forty-five minutes away in Oroville.

"It's a big deal to be here; everything everybody said about coming home, I felt it all. I really feel this place in a way I couldn't feel Fresno," he said. "I wanted to come back with everything and everyone I loved."

As the gym lights dimmed, the chatter quieted, and the players bedded down for the night, Coach Prinz emerged from his office for one last check. Amid all the mattresses and sleeping bags, he noticed one player sprawled flat across the hardwood with his head resting on his backpack. The kid was literally sleeping on the floor. It was Taylor Brady, and the junior receiver didn't look very comfortable. Prinz shuffled up to him, crouched down, and whispered in his ear, "Taylor, how you doing?"

"Sleeping, Coach," Brady replied.

"Where's your sleeping bag?"

"Don't have one anymore."

Prinz scurried away into the darkness and returned with a wrestling mat, an old blanket, and a yellowed pillow that he'd found tucked in a corner of the stage. He dropped them in Taylor's lap and walked away silently.

"The things some of these kids don't have . . ." Prinz said, his voice trailing off into the rhythmic breathing of his sleeping team.

The next morning, Brady awoke with an explanation.

"Our house was destroyed, we lost everything. I didn't have time to buy a sleeping bag, but I was so tired, I didn't really need one," he said. "You've got to understand, I've got to be here. Everyone is watching us. We're setting the example for the town. I couldn't miss this."

As he spoke, his yawning teammates slowly got up, wriggled out of their sleeping bags, and headed to the cafeteria, where awaiting them was another member of the team who didn't sleep much. It was Greg Kiefer, Spencer's father. Greg, the unofficial team cook for the weekend, was functioning on three hours' sleep. Two days ago, he'd showed up in his trailer after forking out $2,000 to buy food from Costco and with a willingness to clean chickens and scramble eggs. With the team requiring four meals to get through a day of camp, Kiefer, a retired sheriff's deputy, had been on his feet for twenty-one hours three days straight, trying to figure out how to feed them. He was part of a large Paradise volunteer army of parents and former players who always seemed to show up when Prinz needed them.

"I'm no cook, but this isn't about cooking, it's about community," Kiefer said. "This is not just for my boy, it's for the town."

As Greg spoke, he realized he would need more water and ice, so he picked up his phone and made a call. In twenty minutes—seriously, twenty minutes—a former Paradise football player walked in with three cases of water and three giant bags of ice.

"Of course he showed up," Kiefer said later. "That's why I love it here."

AFTER BREAKFAST, THE PLAYERS JOGGED down to the field for the traditional end-of-camp ceremonies. There would be a final two-and-a-half-mile run—the "Glory Hill Run," a staple of camp for sixteen years—then they would gather back in the gym to individually stand up and commit themselves to the season. Both acts were considered symbolic of shedding all self-interest and sacrificing for the good of the team. This was where this group of disjointed, displaced kids would officially come together as one.

The 2019 summer run was kicked off by the embodiment of that spirit, forty-seven-year-old assistant coach Nino Pinocchio. Nino wants to spend eternity in Paradise. Literally, that's his plan.

When he dies, he wants his ashes spread across the 50-yard line at Paradise High's Om Wraith Field. He doesn't know if school officials would allow it. He doesn't know if it's even legal. But he does know where he belongs, and it's there on scarred, muddied, hallowed Bobcats turf.

"I came up from that ground, and that's where I want to return," he said.

Pinocchio, stout, loud, and passionately profane, stalked through that ground in a camouflage bucket hat. He was born into Paradise football, grew up with Paradise football, coached Paradise football, and, when the program needed him most, he returned to help rescue Paradise football.

For this 2019 team, he came out of coaching retirement to become a roving coach with no official title and no salary. He did so while holding down his full-time job as superintendent of Butte County Juvenile Hall—and, coincidentally, was Andy Hopper's boss. He also did so while quietly dealing with troublesome diabetes and painful pancreatitis.

Pinocchio was the coach goading the Bobcats' stellar running backs to run harder. He was the coach pushing the Bobcats' quick linebackers to hit harder. He was the coach whose pregame speeches were the most likely to give a kid chills.

"It was all about football, and it was nothing about football," he said. "Every bit of coaching I had learned I had to use, both on the field and off."

He truly came from the Paradise earth, born there to a seventeen-year-old named Cheryl. After his parents separated when he was young, he spent a chunk of his childhood living with her in a tiny ridge home with odd roommates and few frills.

"For the longest time, I never liked meat, because all we ate was canned beef," he remembered.

His father Tony's family owned a Paradise institution called Pinocchio's Italian Seafood, but little Nino wanted nothing to do with the restaurant business. He gravitated instead toward the Paradise pastime of football. Three uncles—Uncle Kevin, Uncle Phil, and Uncle Jim—had all played for Bobcats championship teams. His grandfather John was a referee for thirty-three years. As a five-year-old, Pinocchio would ride with his grandfather to games and learn the rules, eventually memorizing the rule book as he got older. Soon he was watching games his grandfather officiated and thinking, "I can draw up plays better than that." Soon he knew what he wanted to do, and where. Before he ever played the game, he learned football strategy, and he was hooked.

"From the age of seven, I knew I wanted to be a football coach," Nino remembered. "And the only place I wanted to coach was Paradise High."

His mother married a cement contractor named Bill Struve, and, with Pinocchio's two half sisters, they became a family. But this football thing was initially unsettling. It was a problem. His mother didn't want him to play. He was too young. Those Paradise kids were too tough. He begged and begged for permission and was continually denied.

Then, one day when he was at one of his uncle's houses, he devised a plan. His mother had worked the previous graveyard shift at Feather River Hospital, where she was a midwife who'd delivered many of the children in town. It was early afternoon, and he knew she'd be sleeping, so he called her.

"Hey, Mom, can I play football?" he asked her for the umpteenth time.

"Yeah, whatever, I'm sleeping," she replied groggily.

His uncles immediately signed him up for Paradise junior football. That night at dinner, his mother asked him what he had been calling about.

"You told me I could play football!" he said.

"I did?"

Thus began Pinocchio's life as a Bobcat, a brutal yet empowering education at the firm hands of two youth coaches whose names he remembers to this day.

"Ben Di Duca and Scott Sutfin, I'll never forget them," he said. "They set the tone for the Paradise football culture."

The pair, who coached Pinocchio into high school, instilled orderly toughness in this reckless group of mountain kids. They ran. They hit. And if the other team scored, the following week they had to execute "up-downs": an exhausting exercise that entailed dropping to the ground and then jumping back up. Then repeat. One time, they did 120 up-downs, and two kids quit on the spot. The next day, just to prove their resilience, they nailed 121.

"This is where the Paradise players learned that if you had a weak mind, you had a weak heart," Pinocchio remembered. "We may have been a bunch of ragtag kids, but we did not have weak hearts."

Nino was one of the tough guys, estimating that in elementary school alone, he was involved in forty or fifty street fights a year. The kids found a spot on the recess field where teachers couldn't see them, and they would set up a makeshift ring and box. Pinocchio fought one rival five different times.

"We were bored, we lived hard, we'd come out of our nine-hundred-square-foot homes ready to brawl," he remembered. "It was a church pew or barstool kind of town. You were either at one or the other."

Pinocchio came of age during the 1980s, when the Paradise Bobcats were struggling: eight straight years without a winning record. The younger kids saw that, hated the perception of their town as losers, and fought back.

"It sickened me to see how horrible the teams were, yet how tough the kids were," said Pinocchio. "So we got even tougher."

Nino was perhaps too tough. He was kicked out of junior football in the eighth grade for recklessness. His grandfather quit officiating, even after all those years, because he didn't want to have to throw his own grandson out of a game.

"I just did the usual stuff," protested Pinocchio.

He would grab opposing players by the groin. He would attack them at the knees. More than anything, he would tackle with his helmet, spearing them high, taking out quarterbacks with direct shots to the back. *The usual stuff.* Sometimes the other guys fought back, knocked him out, and the next thing Pinocchio knew, somebody was sticking smelling salts under his nose.

Reinstated as a freshman, the young fullback and linebacker led the Bobcats 1987 freshman team to a 10-0 record. He was headed for varsity greatness on both sides of the ball when, in his junior year, he blew out his knee. His senior year, Pinocchio was allowed to play only linebacker, but he would sometimes run into the huddle to sneak his way into the game as a fullback.

"What are you doing here?" his teammates would ask him.

"Shut up and give me the ball," he said.

One year, Nino was given the game ball after an upset 14–0 win over Chico High. His senior year, 1990, the Bobcats finally broke their near decadelong losing streak and finished 8-2. The crowds returned. The buzz was huge. His picture was on the cover of the *Paradise Post*. Life was one big party.

"We would play hard, drink beer, and fight," he said. "It was the Paradise way."

At the time, the Paradise way also meant not looking too far beyond the ridge. After graduation, Pinocchio squandered a chance at getting into a four-year school—he didn't know how or where to apply—and attended Butte Community College instead. Though only five foot nine and 180 pounds, he joined the football team, a venture that lasted about a minute before he tore up his knee again. From there he went to the police academy, then construction, and then hooked on as a sales manager for a juice company in Sacramento. At the same time, Nino began coaching junior football and working as a part-time junior varsity coach in Paradise. He commuted three hours round-trip to Sacramento just for a chance to be on the field. He knew it was where he belonged. He eventually quit the lucrative juice job, took a 50 percent pay cut to work as a counselor and detention officer at Butte County Juvenile Hall, and became the Paradise freshman defensive coordinator in 1999.

Yes, that's the same year Rick Prinz became head coach. And a year later, Paul Orlando joined the program. Nearly twenty years later, these same three coaches would be leading Paradise out of the fire.

"The coaching staff is like the players: our Paradise roots run deep, and we're like a family," said Pinocchio. "Everybody here is connected. That's what gives us our strength."

After coaching the varsity with Prinz from 2002 through 2006, Pinocchio seemingly broke that connection by joining the staff at Butte College as an outside linebackers coach and eventually defensive coordinator. But after helping the Roadrunners win a national title in 2008, and a share of the national title in 2013, coaching almost broke him.

Working both at juvenile hall and on the football team, he regularly put in twenty-hour days, including staying on campus studying film all weekend after games. The stress put him in the hospital three times for pancreatitis. During attacks, it felt like a knife searing through his gut. He finally walked away from all coaching after the 2017 season.

"I had experienced the worst pain I ever felt in my life," he said. "It wasn't good for my health. It wasn't good for my family. I was done."

All of which set him up for the coaching job of his life, on November 8, 2018, when he had to remotely lead his family down a burning mountain.

That morning, retired from coaching responsibilities, Pinocchio was at a hotel in Shell Beach, six hours south of Paradise, at an annual conference of juvenile hall administrators.

His phone buzzed. It was his wife, Valerie. His phone buzzed again. It was his twenty-year-old daughter, Lilly. His phone buzzed again. This time it was a coworker.

"Nino, there's smoke."

"Dad, there's fire."

"Buddy, it's bad. You've got to get your family out of here."

And with that, the coach became a coach again.

"Damn it, I couldn't be there for my family in their time of need. I wasn't present. It's something that will always gnaw at me," he said, his throat thick as he fought back tears. "So, I had to do the next best thing."

That was to give the pep talk of his life. He had to calmly coach his wife, four girls, two sisters, and mother and father out of the gravest of dangers.

First, there was Lilly, who picked up her three younger sisters from school and drove home and called Dad.

"Where do we go? What do we do?" she cried.

"I want you to put a smile on your face, be calm, and get the fuck out of there," he told her.

"Copy, Dad," she said while running to grab personal documents and load them into her Volkswagen Jetta. "I got this."

Next up was his wife, who worked for a local doctor and told him she was calling patients to cancel their appointments.

"You're doing *what*?" he shouted.

Then Valerie said she noticed the bushes around the office were burning; should she call the fire department?

"There's more than a few bushes! The whole damn town is on fire! Get out *now*!" he yelled.

Then his mother called while driving out of town in tears.

"There's flames everywhere! I'm so scared, I have to pee!" she whimpered.

"Relax, you'll get through this. Drive to the juvenile hall in Oroville; they'll take care of you."

Then he talked to his half sisters, Lisa and Cara, and encountered yet another problem. His father wouldn't leave the four-thousand-square-foot home that he and his mother had built from years of hard work.

"Dad says he's not going," said Lisa.

"Then I'm calling one of my probation officer buddies, and he's going to put Dad in handcuffs and drag him out," Pinocchio said sharply.

He was on the verge of making that call when Cara got on the line to say that she'd convinced their father to leave. She wanted to know if she should hook up the family trailer.

"No, no, no!" shouted Pinocchio. "Rush, rush, *rush*!"

Throughout all these conversations, he was speeding up the highway toward home. When he finally stopped for gas in Kettleman City, he slumped over in his car and spent five minutes sobbing.

"I wasn't there," he said in hindsight. "They all needed me, and I wasn't there."

Oh, but he was there, particularly when his mom kept calling him back in increased confusion.

"There's flames everywhere, there's no more landmarks or street signs!" she cried. "What am I driving into?"

Pinocchio calmly directed her out of danger, telling her where to turn, exhorting her with "You can do it, I know you can. Just go, go, go."

The evening ended with his wife and children sleeping in two different cities: Valerie all the way south in Sacramento, the kids in nearby Gridley. Meanwhile, Pinocchio had been summoned to the juvenile hall to oversee a possible evacuation there.

"It was a nightmare that just wouldn't end," he recalled with a shudder.

When they were finally all together again, they lived in three different houses in a span of three weeks before finally settling into a cramped Chico rental. They had lost their Paradise rental home without any insurance. His parents' giant home had also burned.

"We had it real bad," he said.

But one day, after dropping off his daughters at the temporary airport school, Nino visited Prinz in his weight-room office and realized there was somebody who maybe had it worse.

"I saw the kids, I saw how sad they looked, I saw how messed up their young lives had become," Pinocchio remembered. "I thought to myself, 'Maybe if they can bring football back, it will help bring them back, and maybe even bring the town back.'"

So he drove home and told his wife, "I need to coach again."

Then, in March 2019, he left Prinz a voice mail: "Hey, Coach, I need to come home."

Prinz never actually listened to the message until much later, because he knew immediately what it said. He knew his old friend's heart without needing to hear his voice. The minute he saw Pinocchio's name pop up on the voice mail, he called him back to embrace his return.

"We need you," Prinz told him.

"No, to be honest, I need you," said Pinocchio. "This is my program. This is my home. I want to do my part to help us come back. I can't build houses, but I can do this."

Soon after deciding to return, Pinocchio ran into an opposing coach and gave him the news. The coach told him, "I feel so bad for your guys. It's a shame. Paradise football will never be the same. If any of them still want to play, send them to me."

Nino was furious. Then as the schedule was being formed, he heard from other coaches who said they were just happy the Bobcats were going to participate.

"*Participate?*" said Pinocchio. "Those are the fires that burn. It was like, thank you, I appreciate it, now we're going to kick your ass."

The former coach was sold. The inspirational leader was back. When Pinocchio showed up to his first practice, the team was playing on the gravel field next to the airport, and he beamed.

"I thought to myself, 'This is the way it ought to be: Paradise football rebuilding from the rocks,'" he remembered.

His return to coaching was just as rocky, as he would spend his days commuting from Chico to Paradise, to Oroville, and then back to Chico—thirty minutes between each stop—day after day, sometimes feeling too rushed to even take his insulin injections to control his diabetes. Nino put so many miles on his leased Chevy that he had to buy it.

"I felt like a man without a home," he says. "But this was bigger than me."

But he did have a home, and five months after coming out of retirement, on a dew-covered football field on a warming summer morning, Pinocchio took center stage. With Prinz and Pinocchio standing together for their unofficial first handoff of the summer, Prinz instructed the players on the rules of the Glory Hill Run. Then Pinocchio set them off with a story.

In the past, the run would take place on forested trails near one of the community colleges where they held their camp. This year, with the team stuck at home for financial and logistical reasons, the run was to traverse a bike path through piles of Paradise's twisted rubble and rusted metal. It was to be not just a run but also a reminder. This was the route they must take this season. This was the burden they must bear.

"Okay, guys, this is not just about running fast, it's about running together, staying together, doing it as a team," Prinz said. "Now I want you to listen to Nino."

Pinocchio gathered the players around him and told them a version of the legendary story of Spanish conquistador Hernán Cortés and an incident involving his ships during the bloody 1519 conquest of Mexico. Cortés threatened to burn their ships because he thought some of his men were hesitant to fight. Pinocchio exhorted his players to burn their ships so they would begin this journey with the passion that comes from the idea that there is no turning back.

"Cortés said, 'That's how we're going to hold each other accountable: we burn the ships, we go forward, we worry about building them later, we have just one job!'" Pinocchio screamed, spitting the words.

By now, the players knew, that one job was *their* job.

"We're all standing here, we're questioning ourselves, new positions, brand-new starters, are you good enough, do you trust the guy next to you?" Pinocchio exhorted. "Guys, all we have to do is burn your freaking ships! Nothing else matters but the next ten to twelve games! Commit yourself like there's no option of going back, and we'll do absolutely great things!"

Pinocchio slumped over, breathless, and Hopper took over, pleading with the now wide-eyed players to burn those ships.

"You deserve it, and you're going to go earn it right now. Are you going to deserve it and earn it right now?"

The players, on their toes and red-faced from restraining themselves, shouted together for the first time all morning: "Yes, Coach!"

Andy Hopper continued, "Okay, go on this run. Let's be like brothers. Let's freaking go for it!"

"Yes, Coach!

"One, two, three . . ." And here the players shouted from the bottom of their raggedy shoes to the top of their donated T-shirts: *"Burn the ships!"*

For the next thirty minutes, scampering through the charred remains of their hometown, the Paradise Bobcats ran furiously up and down the winding bike path until they returned to the field and prepared to climb the final hill to the parking lot. Once at the bottom of that hill, they waited for everyone to catch up, the slowest joining the fastest, then they split off with a mad dash up the asphalt and past the imaginary finish line marked by the sight of giant Hopper.

"Let's go, let's go! Push, push, push!" he screamed.

Dylan Blood finished first, staggering past Hopper in a sweaty mess, and soon it got messier. The next player finished, stumbled to a nearby chain-link fence, bent over the top, and vomited. Then the next one vomited. Then another one.

Lukas Hartley finished near the front, and he was crying, uncontrollably weeping. Five-foot-eight Jeff Trinchera put his arm around the taller boy and held him up. For years, this has been more than a run, it's been a rite of passage. Even in the best of times, this run was the exercise that weeded out the weak before the season got tough. This was the run where everyone either wins or quits—there's no in-between. The ship-burning talk is tradition. The leaders waiting for the stragglers is tradition. The group vomiting is tradition. And players holding one another up as they walk away is tradition. This run is where the team begins to become a team.

"I've been thinking about the run my whole life," Hartley said. "It's the last time I'm doing it, and I'm leaving it all out there."

After the first finishers stopped throwing up and crying, they started cheering for the late finishers, the slow guys, teammates one and all: "C'mon, Cyrus!" "C'mon, Danny!" "C'mon, Big Dog!" "C'mon, Barrett!"

Oh, yeah, Barrett. He's Barrett Diaz, a wide tackle, and he lumbered across the finish in last place while being cheered and hugged like he was the winner.

"I didn't think I was going to finish it," he said breathlessly as he stumbled away. "I started walking, but the team came back to me and said, 'Run! Run! Run!' Football sucks sometimes," he admitted. "I want to quit." Then he added, "But I can't. I like it too much."

Diaz and his family lost everything in the fire. He moved an hour away to Oroville. He enrolled in school there. He came back in the spring and stayed in a trailer at his sister's house. This run had suddenly become one of his prized possessions.

The last-place finisher was much like the winner. Dylan Blood also lost everything in the fire. He was initially living in a cramped apartment with his family, and, on Glory Hill, he was running from the pain.

"Winning that run pretty much means the world to me," he said. "I know there's a lot of weight on our shoulders. With the town burning down, we want to come back and make everybody proud."

Like Diaz, Blood finished with defiance.

"I couldn't get one thing from my house during the fire—not one, not a single thing," he said, waving his arms at his teammates. "These guys, this team, this is what I got."

The exhausted, exuberant Bobcats gathered around Hopper for a final message before returning to the cafeteria for their end-of-camp commitment ceremony. It was a message that brought them back to the beginning of the run, back to Nino Pinocchio, back to Hernán Cortés the conquistador.

"You've burned the ships!" roared Hopper. "You can't go back! There's one goal. You're gonna win a state title! 'Bobcats will rise' on three! One, two, three . . ."

"Bobcats! Will! Rise!"

THE PLAYERS CLATTERED UP TO the gym for the commitment ceremony, an annual Prinz ritual in which each kid stands in front of the group and states his personal goal, team goal, and character goal for the coming season. They write down these goals on cards, which Prinz saves.

"You never know if there will be an important part of the season where I need to pull those out and remind these guys exactly who they said they want to be," he explained.

The players exhaustedly scattered across the gym floor to listen as their teammates walked to the front of the group and read from their cards. After each recitation, Hartley yelled, "Give 'em three!" and the team clapped hard three times. It was the equivalent of a standing ovation. They were too tired to constantly stand and cheer.

Hartley walked up, still teary from the run, and read, "My goal is to play every practice and every game as a reminder that this will be a day I don't have next year."

"Give 'em three!" *Clap-clap-clap!*

Danny Bettencourt, the quarterback, stood up and said what everyone was thinking: "My team goal is to win a state championship and do it while making history. No one else has gone through what we've been through. We have a bond that no one else can have."

"Give 'em three!" *Clap-clap-clap!*

When the players finished their commitments, the coaches took over the microphone. They'd also filled out cards. In past years, their part in the ceremony had been subdued. But this year was different. This year they turned the simple declarations into sermons.

Shannon Magpusao, a young assistant coach who frequently had fire victims sleeping on his couch, said, "I'm proud to be part of this history that's going to happen. This program is a huge part of rebuilding this community. I take it not as a burden but as an honor."

"Give 'em three!" *Clap-clap-clap!*

Next up was plain-spoken assistant John Wiggins, who'd been coaching at various schools for forty years. He'd now lost two homes in two different fires, and as he hit his stride, the players heard something unusual in his voice: anger.

"The other day somebody said, 'Sorry you're a Camp Fire victim,' and I got pissed off! Victim? No! I'm a *survivor*! I'm just stubborn, and I'm not laying down for nothing. Shit happens. That ball is going take a wrong bounce, and you're going to have to make something happen. The block is not there? You go for the next guy. You miss a tackle at line, you get your butt up and chase that guy down, 'cause that's how you're going to win football games!"

From the floor, Hartley shouted, "Not a victim but a champion!" as his teammates cheered.

"Give 'em three!" *Clap-clap-clap!*

Chip Schuenmeyer, the team trainer and a medic who helped treat fire victims, took the microphone next. Through sobs, he said, "This is our home . . . one day at a time, one step at a time, one battle at a time!"

"Give 'em three!" *Clap-clap-clap!*

Next up, giant Hopper. He was the only one in the room wearing sunglasses, and there was a reason for this. He was preparing to speak his truth, and he wasn't sure that he could get through it. The anticipation in the room was as big as his presence. He stood in front of the group. He lowered his voice. The players sat up from the hardwood. The wheezing old gym went still.

"I've got a little . . . flake in my eye," he said, sticking a giant thumb under those glasses, because he was already beginning to cry.

"November 8 was hard! You lose everything; you think it's the end of the world!" he shouted. "November 13 I got my first opportunity to sneak in. I went to look for my grandma's ashes. I didn't find them. I keep sifting through that, and you start feeling sorry for yourself, and you start thinking nothing is going to be right again. Everything you find . . . you pick it up, it just crumbles, right? All that stuff you're trying to find, it just crumbled. And you start thinking, 'Woe is me, it's never going to be right, it's never going to be the same,' and I gave up. I couldn't find a darn thing."

The players knew the feeling. They'd all been there. They'd returned to their homes in search of trophies and sweatshirts and footballs, and, while sifting through the ashes with a net, they'd caught nothing. The players nodded at Hopper, and some were now crying with him.

"I got pissed off. I threw the freaking netting up, I saw something fly through the air. And I looked over. And it was this . . ."

The coach reached into his pocket and pulled out a blackened piece of metal and held it high. It was a ring: the Paradise sectional championship ring from 2011.

"I bawled like a baby. I didn't know what it was coming out of my eyes," he said, delicately holding the tiny ring in his giant fingers and rubbing it as if it were a priceless jewel. "Look at this: it's all burned and scorched. But when I look at this ring, I think of all those boys that busted their ass every day. And every day I go around town, all those boys come up to me and hug me and say, 'Remember when we did that, Hop?' And I say, 'Hell, yeah!'"

Some of the players were standing now, as if ready to take the field at any moment, fueled by his words, energized by the ring.

"We don't feel sorry for ourselves. There ain't one damn victim in here!" Hopper shouted. "I feel like God chose us. I'm not saying God created that fire; I'm saying God chose us to say, 'You know, I'm going to make these guys the smartest dudes on earth, that they can go through something so horrible and come out the other end and represent to the rest of the world what a man can do.'"

He paused, the players held their collective breath, and then mountainous Andy Hopper let loose.

"You guys want to do that this year?"

"Yes, Coach!"

"You guys want to do that this year?"

"Yes, Coach!"

Hopper handed off the microphone and staggered to the side, tears streaming from beneath his shades.

"Give 'em three!" *Clap-clap-clap!*

The ceremony, and the summer camp, ended with Prinz reminding the players of one of the program's two mottos. It is printed on everything from T-shirts to caps. It is not just a message, it is a mandate.

"We just hit people."

"There's been a lot of changes around here, but we're still going to put on our helmets, put on our shoulder pads, and we're going rock some people, right?" he said.

"Yes, Coach!"

"And if we're a little bit angry about our situation, all the better!" he concluded, in words that portended a season that would be draped in fierce determination.

"We're not the biggest," added assistant coach Bobby Richards, a former player who lost his home and now had to drive ninety minutes each way to help rebuild the team and the town. "But we're going hit you, and we're going to hurt you."

As summer camp ended, the players stalked out of the gym, dragging their new sleeping bags and filled with that now-raging burn. Yes sir, this season they would hit you and hurt you, because Lord knows, they *had* to hit and hurt someone.

"Give 'em three!" *Clap-clap-clap!*

SEVEN

The Opener

AUGUST 23, 2019

Whence with neighbouring arms and
opportune excursion, we may chance reenter Heaven.

The field was ready.

It was the week of the Bobcats' August 23 opener against Williams High, and you could tell that the grass at Om Wraith Field was ready for the return of football just by looking at Bobcats defensive coordinator Paul Orlando.

He was covered in white paint.

Orlando had spent a full day drawing the yard lines as part of his job as school groundskeeper. The field might be weed choked and scruffy, but considering what it had looked like just a month before it was a thing of beauty. When Orlando was first allowed on it, the grass hadn't been watered for eight months and was so badly burned that he tried to raise the funds to pay for new turf. But, under the circumstances, that was out of the question, so, with iron determination and patience, he cultivated the existing grass and fertilized it and slowly brought it back to life.

Orlando shared a thing or two with that grass.

In the Camp Fire, he'd lost not only a home purchased just three weeks earlier but also a successful landscaping business he had spent thirty years growing. Also gone was his father's home, his sister's home, his son's home, and his nephew's home. Oh yeah, and he also saw his horse barn and a chicken coop destroyed. He returned after the fire to sift through the ashes and found nothing.

"Just my family alone, we lost ten homes," he said, still shaking his head in wonder.

Unlike others, Orlando didn't run from the fire at first. With such a large stake in the area, he stayed and tried to fight it with his own water supply. He lost the fight, finally coaxing three of his horses into a trailer and rumbling down the mountain at five thirty in the afternoon, his entire life in flames in the rearview mirror.

"Staying up there that long was stupid, but I didn't think I had a choice," he said in retrospect.

Paul was now living in a trailer on his father's property after initially sharing a trailer with ten people. Before the fire, Orlando's Landscaping serviced fifty clients. But only one of their homes survived, putting him out of business. The school was kind enough to hire him as groundskeeper, but he took a huge pay cut, so that he was barely making ends. He and a couple of other assistant coaches received a stipend of $1,400 a season after taxes, which worked out to less than $1 an hour.

So why didn't he leave? The football. Orlando is a Paradise football lifer. After moving to the town during his senior year of high school, Paul, now fifty-one, stuck around and began coaching at the high school in 1999. The team had since become his second family. In seven years as coach of the junior varsity underneath head coach Hopper, their teams went 67-3, for an extraordinary .957 winning percentage. His two sons had been in the program, one as starting quarterback and the other as first-string middle linebacker.

"Paradise football kept me here and is keeping me going," said Orlando. "It's like we teach the kids: if you get knocked down, you've got to get back up. Football gives us strength. Football gives us normalcy. It teaches us to rebuild and move on."

Paul's landscaping talents had never been more important. The field had become a sanctuary, even if it was initially a dry one. In the early days of practice, when the water purification system for the entire town was still inconsistent, the team had to haul five-gallon water tanks down from campus. Coach Prinz was so worried about keeping his team hydrated, he once snapped at a student trainer who was tardy in delivering water. She cried, and a deeply apologetic Prinz bought her a gift card from Dutch Bros. Coffee, one local business that still stood on Skyway Road. But now, finally, the water was running, and Orlando was painting the lines, and the good feelings were flowing.

Lukas Hartley, standing on the field before the 2019 opener, looked at a reporter's digital recorder and said, "It's a shame that microphone can't pick up these goose bumps," holding up both arms.

The goalposts were ready.

Several members of the Bobcats were taking turns climbing several feet up one of the goalposts and bear-hugging the base, hanging on for dear life. Actually, they were hanging there for thirty seconds—the price to pay for taking off your helmet on the sidelines during practice. Except it was never just thirty seconds. Just ask running back Jeff Trinchera, number 25, who was currently hugging the padded metal and trying to keep from falling.

"Twenty-seven . . . twenty-eight . . . twenty-nine . . . twenty-nine . . . *twenty-nine . . . twenty-nine . . .*" crowed his chortling teammates.

Their laughter sounded strange here. It sounded good. It sounded in stark contrast to the pain that otherwise seeped into their voices. One wanted to cheer for players to absentmindedly remove their helmets and have to climb that post.

One also wanted to cheer for the players to keep their heads on straight. Before the final weekend prior to the home opener, Hartley had a pleading message for his teammates:

"If you drink this weekend, call somebody!"

Prinz quickly interrupted. "How about you don't drink?" he said.

"How about we just go win us a state championship?" interjected Hopper.

The players clapped, but the sound was small. This time last year, the varsity numbered fifty-six. For this year's opener, they would field only thirty-five. But they would take solace in a motto from Hopper's Facebook page:

"One team, one family, one town."

The alumni were ready.

The day before the first game, royalty showed up. He was Jeffrey Maehl, class of 2007, the greatest player to ever wear a Bobcats uniform, and one of those four Division I players to emerge from the program. He starred at the University of Oregon, then went on to play four seasons in the National Football League, spending 2011 and 2012 with the Houston Texans, and 2013 and 2014 with the Philadelphia Eagles. Maehl had come to address the team, but first he spoke about the program.

"You have to understand the mind-set here; the constant, consistent culture," said the thirty-year-old onetime Bobcat star. "They're running the same plays I ran in junior league; it's like I never left. Teams didn't want to play us: they could hit with us for a few plays, but they didn't want to endure it for the whole game. This community was built around that ethic. Going off to college and the NFL showed me what prima donnas looked like, and I wasn't that guy."

Maehl entered a huddle of players. They saw the "Paradise, California" tattoo on his right arm. They knew of the "CMF" tattoo on a hidden part of his body. They were ready to listen.

"I just want to say I'm sorry to all you guys. I want to thank you guys, from the bottom of my heart," he said. "I know it's been tough for you, I can't imagine, but now is the time you take it back. You're a brotherhood. Enjoy the brotherhood. This shit is going to end, football is going to become a fucking business operation, so absorb this, take it all in, fuck the pressure. You were running plays in the dark! I think this game will be over at halftime, but that's just me . . ."

The players erupted in cheers. They needed to hear this. Sure, they talked big, but they had no idea. They'd never played with so few players in such a charged atmosphere. Game over at halftime? In their wildest dreams.

The press was ready.

The night before the first game, a dozen media members gathered around the visiting bleachers. There'd never been this many reporters for an entire Paradise season, never mind a single game. They'd never experienced a press conference like this—in fact, they'd never taken part in a press conference, period. Prinz, Hopper, and seven players filled the metal seats. The reporters scrutinized their weary faces, looking for some sign of the struggle, hoping to capture some sense of the overwhelming emotion. They quickly got it.

Lukas Hartley, is there pressure?

"There's for sure pressure," he said. "Our town is looking for an event, something everyone can rally around. This is the first thing bringing us back together. It's so important, people are telling me Friday is the first day they're coming up here."

Andy Hopper, how are the kids handling the attention?

"We know all the attention is going to go away," he said. "Pretty soon it's just going to be us together. So, it's up to us to create something beautiful and precious, and something generations of folks will tell the story about."

Do you have to win?

"No," said Hopper. "But we're gonna."

Elijah Gould, how are you going to play?

"Everyone on the team has been talking about it forever," said Gould, number 50, a tackle about to begin his senior year. "We're smash-mouth football. We're taking heads. The word of the week is 'Don't come to the mountain.' We love this. We're not going to roll over. We find their weakness. They hate it. We have eleven players who love to hit, and we've been doing this since we were kids."

Lukas, do you want them to feel sorry for you?

"There's a hammer and nail on every play, and we all choose to be hammers," said the unusually poised seventeen-year-old. "I want them to try to take our heads because that's what we're coming out to do. We want no sympathy."

Rick Prinz, what does "We just hit people" mean?

"It's the one message we always have: that we just hit people," he answered. "We're usually the smallest team, our community is smaller than others, so what's the difference? We play football mountain style. We just hit people."

No further questions.

The head coach was ready.

The sky was pitch-black. With the Thursday-night press conference having ended, the Bobcats' season opener against the Williams High Yellowjackets was less than twenty-four hours way. Rick Prinz gathered the team around him for one last pep talk before bed. They dropped to one knee. They listened.

"We started back last January, down in Chico, at the airport," he began. "We had no facilities—we didn't even have a football. You were in your street clothes and cleats. We all went out on the gravel field to run plays, remember? That was a tough time because, in my heart, I didn't even know if we'd have a football team. I didn't know if I'd have a job next year at Paradise High School. It was tough . . ."

Everyone was so quiet, you could hear the crickets in the trees.

"But here's what we did have: we had each other," he continued. "You were scattered and distracted, but when you ran on the field at Marsh, and I saw your faces, you were home, you were happy . . . and we just started moving forward. It was the best spring training I ever had. And everyone started coming back. Lukas went to Oregon, he came back. Ashton went to Fresno, he came back. Junior went to San Diego, he came back. You guys faced so much adversity to get to this first game, you have sacrificed to get to this first game, you have worked your butts off to get to this first game. But to get to this first game is not our goal, is it? What's our goal? To win the game! We're gonna give every ounce of effort in our bodies to win the game!"

The players could no longer remain kneeling or silent. They leaped up and began bouncing around, nine months of anguish about to bubble to the surface against those poor, unsuspecting players from Williams High.

"'Don't come to the mountain,' on three!" Prinz shouted.

"One, two, three, don't come to the mountain!" the players chanted.

Standing behind it all, Hopper just smiled.

"Paradise isn't a town," he said. "It's a feeling."

Blake White's parents were ready.

Brian and Suzanne White, parents of offensive lineman Blake White, were in the stands a day before the opener with a story to tell. Brian, who worked for the California Highway Patrol, had been awake since three thirty in the morning to work the painstaking city job of identifying each of the twenty thousand cars burned in the fire. He was as weary as anyone in town. But he really wanted to tell this story.

"After the fire, I told Blake I'd enroll him in Chico High," he said. "And Blake told me, 'If you enroll me in Chico High, I'll never set one foot on that campus.'"

"So," Suzanne chimed in, "we're renting two apartments in Paradise. Top and bottom. Blake lives in the bottom."

And Blake White still had never set foot in Chico High.

The Dolphinator was ready.

Two weeks before the first game, Randolph Hays, known as Randy, or by his more colorful nickname, "the Dolphinator," showed up to serve as the public address announcer for the team's annual Green and Gold intrasquad scrimmage. He is the father of Bobcats great Jon Hays (class of 2008), who went on to play quarterback for the University of Utah, a Division I school. But that's not why he was there.

"I'm here because right now this is the best place in Paradise!" said Hays, with his hand over the microphone, peering down from the tiny press box at the cluttered field and concrete stands dotted with parents attending the scrimmage. "Seriously, show me a better place in town."

This was coming from a man who had seen lots of places lately. The Camp Fire may have destroyed Hays's home and his lamp shop, but it couldn't wilt his resilience. He lived in eight different places before finally settling in an apartment in Chico.

"This town needs football," he said. "The people need it, the players need it."

As the scrimmage ended and the players and their families hung out in the twilight, Hays had one more message to deliver over the scratchy microphone:

"Great to see everyone! See you in two weeks!" Then a recording of singer-songwriter Phillip Phillips took over, the wistful sounds of his song "Home" playing over the public address system:

"Just know you're not alone, 'cause I'm gonna make this place your home."

Greg Kiefer, one of the parents watching from the stands, couldn't take his eyes off the Paradise kids coming home.

"Watching them hang together, it's euphoric," he said. "Eight months ago, they were all apart. Look at how they've come together."

Standing nearby, Anne Stearns surveyed the scene with the kind of smile few have seen around here in months.

"We're preaching the gospel," said the athletic director. "And the gospel is football."

Their toughness was ready.

There was going to be hitting in a couple of days, there was going to be bruising, and longtime trainer Chip Schuenemeyer was going to be needed. So, Andy Hopper laid out the unique Paradise medical plan.

"If you get an injury and think you shouldn't practice, don't go to Immediate Care, come to Chip," he instructed the team. "You go to Immediate Care, and you're out for two weeks. Chip gives you a high five and never gives you a bill. Immediate Care is for the valley boys. Us mountain boys, we put a little mud on it, a little spit on it, and we go back to work."

Prinz chimed in with his own medical evaluation.

"This is varsity football, and you will be here no matter what," he said. "You could be vomiting your guts out at the time of practice, and that might make you late, but you'll never miss [a game]. You guys are tough kids and tough kids play through it. There's a huge difference between injured and hurt. Of course you're hurt. This ain't going to tickle. What will get you on the field is to be physical."

He finished with a sentence that symbolized a culture: "We don't go through all this to go out on a Friday night to be nice."

So, of course, the kid with the torn knee was ready.

A day before the big game, tackle Elijah Gould hobbled off the field but did not fall. He was playing with a torn ACL in his left knee; an injury suffered in the summer while playing basketball. He was going to need surgery, but doctors said it couldn't get any worse, so he decided to play until he dropped.

"I'm going to be running on it lighter than I normally would, but I'm going to keep moving," explained Gould, who stood five eleven and weighed a solid two hundred pounds. "I have to keep moving. This is going to be our rebuild season, all eyes on us. This is our little town, and I want to be here for it."

Another indication of his motivation can be found in the three English letters tattooed on his left bicep. You know what they are. Since he wasn't eighteen yet, Elijah had to drive all the way to Reno, Nevada, a half hour away, and have it inked there.

"CMF is a big tradition in this town," he said. "I'm just carrying on that tradition."

The parking lot locker room was ready.

In keeping with the occasionally unsightly weekly tradition, the Paradise players dressed for the opener in the gym parking lot. They stood on their flatbeds to squeeze into their pants and pads, they sprawled out on the concrete to fit into their shoes, and then they basically roamed the blacktop in their underwear for thirty minutes while fans walked past with a shrug.

"I tell my bosses, 'If you want us to change in the locker room, give us a locker room,'" Prinz said with a shrug.

The marching band was ready.

They weren't going to march. Indeed, they were barely a band. But nineteen Paradise musicians—about half the size of last year's squad—showed up for the opener anyway, blowing and pounding and playing their hearts out. They had three tubas but just one trumpet. And one of the drummers was longtime band instructor Bob Schofield. Yet they sat squeezed into a couple of bleachers and played on.

"We will forge ahead," announced Schofield, echoing the mantra of thousands.

The paramedics were ready.

The approximately five thousand fans who crammed into the bleachers at Om Wraith Field for the early evening kickoff—a couple thousand more than normal—were startled by one of the fire's giant fingerprints. Because so many surrounding trees had burned, the setting sun was free to pound the spectator seats in the ninety-five-degree heat and turn the place into a dangerous sauna. At least four fans were treated for heat exhaustion before the game even began. Still, nobody moved. They'd all experienced hotter.

"Tough town," noted Greg Kiefer.

The snack bar was ready.

Six bucks for nachos. Five bucks for a burger. Seven bucks for a double burger. A buck for a soda, Gatorade, water.

And one weeping Richard Van Stavern to serve you.

Van Stavern, seventy-five, graduated from Paradise High in 1962, lost his home in the fire, and now had shown up four hours early to help with the rebirth.

"This is such a shot in the arm," he said from his spot behind the cluttered snack bar counter, tears filling his eyes. "Look at us, back on our own field again. Makes this town very happy."

The town was ready.

Four hours before the game, they began showing up. Folks with wonder in their eyes. Folks with tears streaming down their faces. They walked up to the field as if greeting a long-lost friend, pointing, giggling, crying.

The bathrooms weren't working, so there were lines at the Porta-Potties. Folks were standing shoulder to shoulder behind one end zone because this was the biggest crowd in school history, and there was barely room for everybody. The stands were full. Their hearts were full. Everyone hugged, and nobody complained.

"This day is bigger than football, it's about our community," said Jeff Marcus, the new (and old) principal of Paradise High. After Loren Lighthall left to take a job out of town, Marcus, who'd held that position previously for five years, came out of retirement to both run the school and coach the kickers even though he'd lost everything in the fire. "It's a time of healing, reuniting, moving forward together."

In all, Marcus had been a principal in the school district for twenty-three years. Before coming to Paradise High (the first time), he'd spent eighteen years at nearby Ridgeview High School, in Magalia. His daughters had gone to school here, and now his grandchildren were

Paradise students. His wife was born and raised here. He was temporarily living on a rice farm thirty minutes away, but this would forever be his home.

"I can't imagine somebody coming from the outside who didn't understand the community," he said. "It's a different culture. It's Bobcats culture."

The first member of the community to show up for the game arrived more than four hours early: lone tailgater Michael Weldon, the father of running back Ben Weldon and former Paradise receiver and running back Jacob Weldon. They'd lost everything in the fire. Michael's Toyota Tundra pickup was new, as were his green T-shirt, shorts, tennis shoes, and sunglasses. He worked nights sorting mail, and he was exhausted from sorting out his life, but didn't want to miss this moment for his son and his town.

"The only photos we have left are the memories that pop up on Facebook," he said sadly. "We need to make some new ones. This game will be the most important game in our lives. This game will be everything."

Soon another tailgater arrived in an adjoining vacant lot that once held a church. Matt Madden, a Chico police officer who used to coach in Paradise, threw up a tent, fired up a grill, and, even though he'd extended no formal invitations, waited expectantly for folks to show up. He was not disappointed. A large crowd began to fill up the parking lot, many of them displaced friends who were seeing one another for the first time since the fire. There were hugs and tears and prayers.

"My house survived, but everyone around me is gone, all my friends gone, nobody is left, and to see this today . . ." Madden said through tears. "Everyone coming back now, believing this can be a town again, it's really something."

Everyone eventually walked down a narrow paved road to the field. Standing behind one end zone was former mayor Steve "Woody" Culleton, a local institution. He was crying, too. "People are finally coming home," he said. "We lost our community, and today we're getting it back."

The pregame prayer was ready.

Thirty minutes until kickoff. The Bobcats players and coaches gathered in the gym above the field. They took a knee. They placed their hands on one another's pads. They bowed their heads.

Linebacker Josh Alvies led them in prayer.

"Thanks for bringing us together," he said softly. "Thanks for bringing us back to Paradise."

Alvies said he had planned a different prayer, but something else came out. He said it was his voice but clearly God's inspiration.

"I wanted to emphasize that the fire could not stop us," he remembered later. "I wanted to pray for our brotherhood."

The prayer completed, the Bobcats rose slowly and prepared to line up and head down to the most important game of their lives when Prinz stopped them. He had one last word.

"They thought you were down, they thought you were weak, they thought you can't fight back, and you know what I say to that?" Prinz screamed. "Hell, no! Lay it on them!"

They gathered in a huddle one last time, stuck their young hands together, and shouted loud enough to be heard outside by a Yellowjackets team that had no idea what was about to hit it.

"Don't Come to the Mountain!"

Johnny Cash was ready.

For more than a decade, before every home game, the Paradise Bobcats had marched down through the stands and onto the field accompanied by Johnny Cash, whose "God's Gonna Cut You Down" blared over the PA. It was a dark, traditional American folk song with a toughness that appealed to Prinz when he started the tradition.

On August 23 those words and that tradition carried an entirely new meaning for a football team that God had clearly cut down:

You can run on for a long time
Run on for a long time
Run on for a long time
Sooner or later, God'll cut you down
Sooner or later, God'll cut you down.

Prinz always said it was a good football song. Now it was something more.

"It's literally what happened to us," he said. "Now we'll use the song to climb back up."

They marched down together with both hope and remorse, wearing dark green helmets donated by Chico native and Green Bay Packers star Aaron Rodgers, cleats donated by Under Armour, and black jerseys purchased with insurance money. For this first game, they were led by thirteen recent graduates whose senior home finale had been stolen by the fire. Gretchen Cassing was not among them. The sting of the lost season was still too much to bear. But others came to grit their teeth and remember.

"They didn't know that last year's last game was going to be their last time here," said quarterback Danny Bettencourt, number 5, who would be taking over as starter from Colby Cline. "We're brokenhearted for them. They deserve a chance to do it again."

The graduates gathered in the end zone as if preparing for one last game. This march had never meant more. God had cut them down at the end of a dream, and this was their chance to experience that dream once more, albeit from the sidelines.

"Nine months ago, something was stripped from us. Everything was lined up; next thing you know, everything is gone. We never got our last game," said former lineman Ezra Gonzalas. "To have it all ripped away and to be able to return and carry the flag down . . . this is closure."

It turned out to be a loud and banging closure. When the players made the right turn at the press box and headed down through the stands, they saw a newly erected metal plaque memorializing their nightmare. "C.M.F. 11-8-2018," it read. The players instinctively stuck out their hands and hit it, one open-handed smack per player, serenaded by the Man in Black and a standing ovation that produced chills.

"I felt like hitting it was the right thing to do," said Bettencourt. "We're always going to remember it, but we're also trying to move past it to create new memories."

Their march ended on the field, where they stood solemnly as a student sang the national anthem. In the middle of the song, the microphone went dead, but the crowd, filling the stands and lining the fences, just kept singing. It was then that giant Andy Hopper broke down, tears streaming down his grizzled cheeks.

"I was thinking, we lost everything, but maybe now we found it again," he said.

The Paradise Bobcats were ready.

Oh, yeah, they found it again. They found it quick.

They struck on their first possession, with Hartley running over a Williams kid on his way to an 11-yard touchdown. He was still so nervous, though, that on the Bobcats' next possession, he threw up on the field.

"The puking felt real," said Hartley later, sweating under an American flag bandanna. "Everything else felt like a dream."

They struck again on their next possession, with a 64-yard pass play from Bettencourt to wide receiver Mason Cowan, number 19, who, despite weighing just 140 pounds, tossed aside a Williams tackler before springing into the end zone. His teammates swarmed him like he was a hero, but Cowan knew real heroism. Nearly a year earlier, he lived it.

ON NOVEMBER 8, 2018, COWAN, then a junior, was walking to school with his twin brother, Gabriel—skipping, really, because that's how eager he was to hang out with his teammates the day before the Bob-cats' playoff game. He played for the camaraderie. He longed for the togetherness. He felt like he had spent his entire childhood waiting for that moment.

The five-foot-ten, 140-pound receiver, who was the slightest kid on the team, moved to Paradise when he was nine but didn't play football until high school. He could never afford it. Once he joined the team as a freshman, with equipment supplied mostly by the team, nobody appreci-ated the opportunity more. Mason was always considered small, but the Bobcats toughened him into something big. He'd always looked at the football team like a kid with his face pressed against a Christmas window, but now that he was inside, he felt enveloped in the trademark Bobcats warmth.

"I got there and could instantly tell that everybody was close," he said. "It was a really rich environment."

It was rich in everything but wide receivers. Rick Prinz's traditional wing-T attack was almost exclusively about running the football, so wide receiver wasn't considered an important (or desirable) position.

Cowan set out to change that. His freshman year, not a single pass came his way. But as a sophomore, he became a starter on the junior varsity, and by the time the team reached the playoffs during his junior season on the varsity squad, in 2018, he was ensconced in the tight co-coon of the Bobcats' culture.

"I couldn't wait for those playoffs," he remembered. "Catching a big pass in a big game was something I dreamed about. It was going to be like something you see in the movies. It felt like such an honor to be in that spot. I was made for something like that."

Walking to school the morning of November 8, Cowan soon learned he was made for something else: saving lives.

As he and his brother approached the Paradise High campus, they were met not by his football family but by teachers pointing to the blackening sky above them.

"No school today!" one shouted. "Get your parents and leave town!"

For the twins, there was one problem with that directive: their parents had already driven off to their jobs in Chico, leaving Mason and Gabriel no easy way out. They called their stepmother and father, arranged to meet them back at their house, and began to walk home.

Soon they were running. The twenty-minute trip was now decorated in falling ashes and toppling trees and burning cars, and the boys were racing for their lives. They were sprinting around embers and through smoke when suddenly Cowan received a text from his girlfriend, Anabel Lois.

Her aunt lived in an apartment in a retirement community located on Cowan's route home. The aunt had not responded to calls and texts, and Lois was worried that she wasn't aware of the incoming danger. Could Mason please stop by her aunt's house, bang on the door, and tell her to leave?

"I got this," Cowan assured her.

He raced to the aunt's front door, wailed on it with his fist, and an older woman in her nightclothes opened up.

"There's a fire out here!" said Cowan frantically. "You've really got to get out of here."

As she thanked him and closed the door, Mason noticed that the surrounding apartments seemed quiet. By now, ash and burned bark were falling on his shoulders. He didn't have much time. But there were human beings behind those doors; older people who might have no idea they were on the verge of being burned alive.

Mason Cowan figured he'd have to make time.

He banged on another door, and another elderly lady, also in her pajamas, appeared with confusion in her eyes.

"Ma'am, you have to get out. Pack your stuff, please," he pleaded.

Then he banged on yet another door, and an elderly man answered while toweling off after a shower and rubbing his eyes.

"Sir, this fire is serious, you have to evacuate now!" he said before sprinting down a hallway.

One more door. Three more bangs. Another frazzled senior citizen opening up with a surprised expression. "Go! Go! Go!" he shouted.

Then Cowan was gone, now sprinting home with Gabriel around traffic-stalled cars filled with crying families. He realized he'd risked their lives by stopping at those apartments. But he also knew he didn't have a choice.

"There were people in danger, somebody had to help them, and, at that moment, it had to be me," he remembered. "Looking back, it was the dumbest, best thing I could have ever done."

He and his brother met their father and stepmother at the house. The two boys climbed in the car with their stepmom and sped off while their father, Shane, stayed behind to clean pine needles off the roof in a futile attempt to save the house. He soon left, too.

Driving down the mountain, Cowan looked at the flames surrounding their car and realized that even if they made it, their father, stuck on a side street far behind them, might not. "I might have to leave my car and try to run out," Shane told them. "I love you and—"

The cellphone connection went dead. Cowan finally broke down in tears. The sky was black. His dad was stuck in flaming traffic. Mason, Gabriel, and their stepmother eventually drove out of the burning village and down to Chico, but still no word from Shane.

However, like son, like father.

"I knew my dad and how he raised us. I had that tough skin, and so did he, and I knew he would drive through a wall of flames to get to us," recalled Cowan.

Shane did just that and arrived in Chico to a tearful reunion. There was also the good news that Anabel Lois's aunt had escaped the fire, too. As far as Cowan knew, so did the residents of the three other apartments that he'd alerted.

Eventually the Cowan family moved to their grandparents' house in Oroville. Once there, setting up camp on a living room couch that would become his home, Cowan began conspiring with teammates to beg Prinz to let them play that postponed playoff game.

"My house had burned, I was physically and emotionally depressed, I needed football—we all needed football—so we told Coach Prinz we would do anything to play that next game," he said. "We would wear other teams' gear. We would play on their fields. We would dress in their locker rooms. We didn't care."

It wasn't happening. The season was done. The hurdles were too high, the strain too great. Mason Cowan's dreams of being a postseason hero were dashed, and he fell into a deeper funk.

"Every day I woke up, and the realness of life hit you. Every day that passed, I was more depressed," he reflected. "I vowed if I got another chance the following season, I was going to make a difference."

On the Bobcats' second possession of the 2019 season, he got that chance, and 64 yards later, he had scored the touchdown of a lifetime.

"I get the ball, my heart was racing, I juke one guy, I run over another guy, I get into the end zone, my heart is still racing," Cowan said, still excited by the memory. "I'd dreamed of this moment. I was made for this moment. My whole life was for this moment."

WITH COWAN'S TOUCHDOWN DOUBLING THE score to 14–0, the Bobcats' poor opponents were overwhelmed. Remember, Williams High was initially offered $5,000 in equipment from Under Armour to play this no-win game. The school didn't accept it, but suddenly that figure seemed low.

Before the game, Williams announced that its students were donating a week's worth of lunch money to the Paradise students. The generosity was deeply appreciated, though, at the moment, the Bobcats didn't look like they needed any more fuel. The players hit so hard, you could hear the pads crash. They chased so furiously, you could hear the Yellowjackets players gasping for air. The Bobcats ran over them, around them, and usually ended up on top of them. There was a moment in the first half when the sunset peeking through the trees resembled a giant ball of fire. But this time it burned with beauty. This time the town of Paradise was glowing.

The Bobcats led 21–0 after one quarter and 35–0 at halftime, at which point they were rudely reminded of their challenges when they repaired to what must have been the worst halftime locker room in football history. The players congregated above one end zone on a burned-out space of weeds and rock. There were no permanent lights. There were no restrooms. There was nowhere to even sit. It would take too long to climb up to the gym and use its facilities, so this was their halftime retreat, and Prinz wouldn't hear of the discomfort.

"Let's keep pounding!" he shouted.

"Get off our mountain!" they shouted.

The second half was just as furious, with Paradise rolling to a 42–0 victory that wasn't even that close. Jeff Lemus, in only his third week as head coach of a Williams team that was physically smaller and less skilled, trudged off the field like he knew he never had a chance. This was originally scheduled as a relatively even match against a fire-torn team, but Paradise didn't play like a fire-torn team. They played like a team on fire.

"It was just a hard situation," Lemus said. "There was a lot of emotion out there."

When the game ended, that emotion left Bobcats lineman Silas Carter, wearing number 32, writhing on the ground with leg cramps as his teammates danced around him.

"Hell yeah! Hell yeah! Hell yeah!" they chanted.

Andy Hopper nodded at the kids with a sweaty grin.

"Can you feel it?" he asked. "Tonight the healing began."

Nino Pinocchio interrupted their partying to point to the concrete stands, which were still filled with bouncing, hugging fans.

"Look around you! There's a helluva lot of smiling faces up there!" he screamed. "You did that! *You did that!*"

Prinz waved the crowd onto the field, where team and town formed a big circle, everyone hugging and leaping, grandfathers and children, mothers and sons, helmets and pads and heart—seemingly all of Paradise—waving fists in the air in delight and defiance.

"We're going to cheer together!" screamed Prinz, and so they did, chorusing three letters that described a culture, three letters that marked the beginning of a journey, three letters that the flames could not touch.

"CMF! CMF! CMF!"

The Bobcats were 1-0, and the mountain folk were going crazy.

EIGHT

Hopper

AUGUST 30, 2019

*The law of faith, Working through love, upon their
hearts shall write, to guide them in all truth.*

For some on the Paradise Bobcats, the fire lived in their heads, leading them to the football field to seek respite.

"It's the only thing that has any importance in our lives," said Lukas Hartley.

For others, the fire lived in their hearts, pushing them to recklessly attack their football opponents as if their foes were holding a gas can and matches.

"We're way more focused, we're not messing around, we act like the other team took our homes from us instead of that fire," said running back and safety J. D. Webster.

For 370-pound coach Andy Hopper, the fire simmered and stewed and baked in his belly. The Sunday after game two, it exploded.

Hopper, forty-seven, was the team's emotional touchstone. He was not only their offensive line coach but also their connection: the one who screamed at them, joked with them, cried with them. This emotional behemoth knew how to reach those crazy mountain folk because he was one of them.

"He is the one who really speaks to us," said Hartley.

Hopper grew up on the hill, in the rural burg of Magalia, just outside of Paradise, with his mother in what was essentially a two-bedroom garage. His father, Bert, died when Hopper was six, but the boy never knew him because Bert had spent most of those years incarcerated for selling heroin. Officials released his father from jail when he was dying of cancer, leading to an unusual farewell between father and son. Although Bert was only forty-seven, the drugs had aged him, and he was withered, with gray hair and beard.

"Boy, do you know who I am?" asked the father.

"Santa Claus?" said the son.

"That's right."

So Hopper went through much of his childhood believing that his father was Santa Claus. He had two photos of his father. Both were lost in the fire.

Hopper spent much of his childhood hanging out at the Optimo Casino Club, a local dive bar and restaurant owned by his grandfather Red Hyatt. There was a card room in back, hookers in front, and Hopper hanging out with everyone, eating prawns in the kitchen and maraschino cherries from the bar while doing his homework in the storage room.

"I loved being there; it taught me that everybody is worth something," he said fondly. "Besides, I could never be left at home. I'd eat everything in sight!"

Hopper grew up with a collection of mountain buddies, running wild, no direction. The back roads were their home, the laws their own. He sported a green Mohawk haircut and a hyperactive nature. He sewed his own clothes or wore Kmart T-shirts over spandex pants. He and his buddies fought their own wars. Somebody's alcoholic father beat up somebody's mother? Hopper's gang would rough up the guy. Racing up the mountain lanes? Check. Disappearing for a few days in the woods to hunt and fish, even as young teenagers? Check. Getting thrown off the school bus for crawling under the seats and tickling girls' feet? Check.

"We weren't looking for trouble," Hopper said with a good-natured laugh. "We *lived* in trouble."

When Hopper was thirteen, his stepfather threatened to kick his dog, and Andy fought back. His stepfather got him on the ground, at which point his mother, Beverly Jean, ran out and put a whopping on his stepfather.

"Yeah," said Hopper, "we lived in that kind of trouble."

Seeking an outlet for his wild side, Hopper naturally joined the Paradise High football team as a freshman. He had been too big to play youth football, so this was his first experience at the game. When a coach told him to block, he tackled. He played center and defensive end his freshman year, but something bigger happened that season. That is when he met the feisty new youth minister at Magalia Community Church named Rick Prinz. Hopper and his gang had been adopted by the church when they were caught breaking windows there in sixth grade. Since then, he saw the place as a quiet refuge from his crazy life, and Prinz soon became one of his saviors. Where others regarded Hopper as a raucous brute, Prinz saw him as a sensitive soul who constantly had his feelings hurt and who reacted with humor or tears. Where others condemned Hopper, Prinz embraced him, dirty feet and all. Hopper, you see, never wore shoes.

"As soon as I'd walk into their house, Veronica would make me go to the bathroom and wash my feet," Hopper remembered. "I learned religion was just people who loved and cared."

After playing fullback his sophomore year—and being benched because he could never hold on to the ball—Hopper quit before his junior year. He had no choice. His mother lost her job at the bank, and he couldn't even afford cleats. He spent the year working at an RV park while being nagged by one particular Paradise coach.

"You've got come back and play football," Prinz would tell him. "You've got to stay engaged by playing football."

Prinz knew that Hopper needed the structure. This was a kid who once got his head stuck between the metal bars of a fence; the fire department had to be summoned to extract him. This was a kid who would spend two minutes on an economics test and turn it in blank while his classmates howled. Early in his senior year, he asked a teacher if he was going to graduate, and the teacher replied, "Yeah, because there's no way in hell I'd have you back again."

"I would do anything for a laugh because I felt like an outsider all the time," Hopper recalled. "I loved to make people laugh, love to make them feel good about themselves, because I know what it's like not to feel good about myself."

His mother started working again, so Hopper returned to the Paradise football team for his senior season while toiling at a local Mexican restaurant. By now, Prinz was a varsity assistant, so they were together on a football field for the first time. Hopper lasted two games. He hurt his knee, didn't have the money for surgery, and his playing career was done. But Prinz wasn't done with him.

"You can stop playing football, but you can't quit on me," Rick told him. "You are a leader of leaders. Lead in our youth group. Lead in our community."

Today Prinz contends he's never been around a more magnetic personality.

"The kids all followed him, like a pied piper in a green Mohawk and flip-flops," he recalled. "It's his charisma. It's a gift."

After three years at Butte Community College, Hopper dropped out and started teaching special education. He thrived working with disabled children. He also began coaching the freshmen at Paradise High School. Eventually he became head coach of the junior varsity team, which, along with defensive coordinator Paul Orlando, he helped lead to a three-year unbeaten streak. He stressed to the kids that, like him, they were all outsiders—mountain folk that nobody understood.

"I would tell them, when we played all these bigger teams, we all grew up with nothing, now is our chance to take it from them," Hopper said, and part of the Bobcats' culture was born.

He inspired them with a 1993 song called "Hyena" by the punk-rock band Rancid, with one of the lines being, "I'm a hyena fighting for the lion's share."

He would tell his kids, "We're not the lion, we're the hyena. We can't go it alone, we're too small, and our teeth aren't big enough. But when you have eleven hyenas, you can surround the lion and steal his meat."

Hopper moved up to assistant varsity coach in 2008, and had been there ever since, rebuffing offers to coach in bigger programs because he believes his purpose is of a higher calling than football.

"The reason I coach is to reach Paradise kids, because I am one," he said. "I don't fit the mold of any other school."

To earn a living, Andy took another job with rambunctious children, spending fifteen years as a counselor at Butte Juvenile Hall, doing everything from mopping floors to restraint. He'd seen massive fights, a kid jumping off a second-story guard rail and landing headfirst on a table, kids breaking other kids' jaws. But he'd also talked kids out of crisis, talked them down off ledges, talked so they listened. It was this experience that, later on, helped him deal with troubled Paradise players. Hopper counseled the player whose dad made him cut down his dead brother who'd hung himself from a tree. He comforted the player who missed practice because he'd just buried his horse. There are players who lived with him because they had nowhere else to go.

"When I walk onto the field with those kids, I am those kids," he said.

His first day on the job at juvenile hall, Andy Hopper knew he could have an impact on young men. He encountered a naked kid slamming a broken mop handle against a door, screaming, "Fuck you, fuck you, I'll kill every one of you!"

The kid saw the giant coach, remembered him from church youth group, and began crying.

"Hopper! Hopper! Hopper!" he screamed.

THE BOBCATS WERE COMING OFF a 35–28 win against the East Nicolaus High Spartans, a suitable encore to their opening victory over Williams. Played in the tiny farm town of East Nicolaus, population 225, it was a game filled with mosquitos and a message about the Bobcats' mission. Before the kick, East Nicolaus officials asked Prinz if he wanted some sort of memorial ceremony, much like when Williams High School donated a week's worth of lunch money.

"Nah," said Prinz. "We just wanna play."

Adorned in white jerseys with green helmets, play they did, running for 451 yards in a game that was close until Hartley barreled for touchdown runs of 2 and 32 yards in the fourth quarter, and the defense held to preserve their second victory in a row. It was such a tough game that Spencer Kiefer, carried off the field on a stretcher early on due to a knee injury, refused to stay on the sidelines. Number 1 limped back out and played the rest of the game, screaming, "I'm not giving up unless I'm in a wheelchair!"

Afterward, upon returning to Paradise, instead of celebrating, Prinz and his coaching staff were filled with worry. In past years, after returning from road games, the kids couldn't wait to scamper from the bus to their cars, so they could speed out of the parking lot on the way to whichever player's home was hosting the party. But on this night, they stuck around, sitting on the hoods of their cars or in their front seats, all revved up and nowhere to go in a town that no longer existed.

"Go home!" shouted Prinz, fearful of the trouble that could arise from a bunch of teenage boys loitering in a parking lot.

Yet for most, they *were* home.

"There's no place left in Paradise to really hang out; you can't go to anybody's house, so the parking lot becomes our entire social scene," said Hartley, who had just rushed for 277 yards.

Prinz understood it. He just didn't know what to do about it. "The problem is," he said, "many of them feel like they have nowhere to go."

Andy Hopper saw this and was deeply bothered. The weight of maintaining the connection to the kids when everything around them was broken finally proved too much, even for him.

The following Sunday, Hopper awoke around eleven at night feeling like his insides were on fire. He was rushed to Enloe Medical Center in Chico and then placed on a helicopter for a hospital in Sacramento. But the helicopter couldn't land there due to fog, so it was rerouted to Renown Regional Medical Center in Reno. The diagnosis was an aortic dissection: a tear in the inner layer of his aorta, the trunk-like arterial vessel that receives blood from the heart to be distributed throughout the body. He was given a 6 percent chance of survival.

Prinz was devastated. Andy was his wingman. His best friend and confidant. How could he coach this team without him? What if Hopper died? Surely he wouldn't die. The world wouldn't actually do this to all those who loved him, would it? After ruining their town, it was going to take their leader, too? Rick knew he had to be stronger than ever, but he didn't know how. A day after Hopper's hospitalization, Prinz tearfully broke the news to his players during their sixth-period weight-training class.

"He may not be back," Prinz told them. "If you pray, you should be praying for him."

He looked at their young faces and saw the strangest thing. Nothing. No emotion. No tears. No wailing. Nothing.

"They had this glazed look on their face, like this was just one more thing, one more loss, and they couldn't take anymore," Prinz recalled. "They just stared at me. It was the saddest look in the world."

The Bobcats were 2-0, but the road out of hell just got rougher.

NINE

The Senator

SEPTEMBER 6, 2019

Here on Earth, God hath dispensed
his bounties as in Heaven.

It was the third game of the season, against smaller Gridley High School, and Rick Prinz was lost. He was coaching horribly. He couldn't stop thinking about Andy Hopper, his giant sidekick who was still clinging to life in a Reno hospital. Prinz thought he should be there with him. He thought he should be in church praying for him. He thought he should be anywhere but on the sidelines, trying to lead his own group of lost souls.

"My head wasn't in it," he admitted later. "My heart wasn't in it."

This time it was the Bobcats who saved him. In the September 6 home contest against the Gridley Bulldogs, Prinz basically coached for one quarter before losing all focus, but it was all the coaching they required, as they raced to a 21–0 lead in those first fifteen minutes and cruised the rest of the way to a 28–6 victory.

"I did a really bad job as a play caller," Prinz recounted. "I just wanted to go ahead and get the game over with. I was crying the whole time. Tried to coach the game and just couldn't do it."

How much was Prinz out of sorts? He didn't call one pass play the entire game, which was unusual even for his ground-heavy offense. The

Bobcats were saved by 203 rushing yards from Lukas Hartley and another 105 yards on the ground from Tyler Harrison. It was becoming obvious, this unlikely duo who both returned to Paradise without their parents was going to be a formidable force indeed. The previous year Hartley had played in just one varsity game, Harrison played in none. They were very different runners—Hartley pounded and Harrison sprinted. Few had any idea they would work so well together, and few believed they could outrun the previous year's star duo of Dominic Wiggins and Jacob Weldon. But from the beginning, they were nearly unstoppable.

And against East Nicolaus they were carried, and not for the last time, by the plowing blocks from a beefy offensive tackle named Kasten Ortiz, who repeatedly cleared the path for Hartley and Harrison. He pushed past the insistent fear for his line coach's life. Wearing an American flag bandanna and a determined grimace, the six-foot-two, 220-pound senior settled this most unsettled team. After all, Ortiz was always good at hiding the pain.

As a child, his stepfather, Chris Crowley, would whip Kasten with a belt, and he wouldn't flinch. His head would get smashed against furniture, and he would shake it off. He would get slammed against a dresser, and the boy would walk away. He had bruises, welts, and nobody saw.

"I would wear baggy clothes," said Ortiz of his childhood in the central California town of Orcutt. "Not a visible mark on me."

He grew up not only abused but also confused. He was ten before he realized that Crowley was not his real father. Then, at age twelve, following Crowley's death due to a heart condition, his mother, D'Ette, was struggling with substance abuse, so Kasten was placed in a foster home. At that point, he didn't know where he fit.

"I was always looking for a sense of belonging," he said, displaying a degree of self-awareness that belied his young age.

Then one day in the summer of his twelfth year, he was summoned to the phone.

"Hey, Kasten, I'm your father," said Bobby Ortiz.

Kasten was both stunned and delighted.

"Oh my God, I really have a father," he said.

A few weeks later, the youngster received another surprise when he was told to pack his things: his father was coming to pick him up.

"Where are we going?" he asked.

"To a place called Paradise," he was told.

And so the changing of another life was under way—yet another refugee from an unsettled situation discovering himself among the tall trees and deep roots of the ridge. Two days before the start of sixth grade, twelve-year-old Kasten Ortiz moved into his new home.

"I smelled the air, I saw the trees, I was entranced by the environment," he said. "I felt like I had been saved."

He began playing football in the summer before seventh grade, and, slowly, an unsure, chubby five-foot-two, 160-pound kid grew into a determined six-foot-three, 200-pound leader.

"I had never played football before, I had no idea what to do, but it was thrilling. It gave me a sense of belonging and put a lot of my unrest to rest," he said.

Still, he wasn't certain he belonged. He became the goofy kid in the back of the room, the one who would shout outrageous jokes, the class clown. Except he wasn't always funny. And he *knew* he wasn't always funny.

"I'm not a fighter, I want to make people laugh," Kasten remembered. "But I tried too hard. I could be disrespectful, I hurt people. It was shameful."

This shame reached a low point in his sophomore year, when he earned a one-day suspension for making a joke about a student's yarmulke. He wrote a letter of apology. He still can't forget the pain he caused.

"I was so hurtful, that insensitive act will live with me forever," Kasten reflected. "I realized that is not the kind of person I am. I had to be better."

Finally, he stopped trying so hard to be a comedian. Soon he learned he could get all the attention he wanted on the football field.

"I finally found a place where I could get everything out," he said. "I spent so much of my life burying or masking my emotions, I could finally let it all loose on the offensive line."

His grandmother's house, where he had been living adjacent to his father's home, survived the fire. Ortiz promised himself he would repay his good fortune by becoming a team leader in his 2019 senior season. Three games in, he was already fulfilling that promise.

"Kasten was our hardest worker, our most consistent, faithful to his word, and fully committed," said Prinz.

He won a starting job at tackle and immediately became a cornerstone. Ortiz began tutoring teammates who struggled academically. Number 58 was always one of the last players off the field after practice because he would be lugging in the blocking bags. He became part of Paradise's Link Crew, a leadership group that helped freshmen acclimate to the high school environment. During the team's trip to the 49ers game immediately after the fire, he gave his sweatshirt to a shivering teammate who had none. Once, Prinz even used him to vet a pep talk. He was like the team's senator.

"I'm giving everything I have in my senior year," Kasten said. "I'm leaving it all out there."

When the Gridley game ended, Ortiz hung out in the end zone long after the crowds had departed, chatting with family and friends and soaking in the miracle of growth.

Looking back on the 2019 season, Ortiz said, "That year means everything to me. It is very impactful on who I am as a person. It helps me realize how important and special every moment is."

The Bobcats were 3-0, and the moment was alive.

TEN

The Preacher

SEPTEMBER 13, 2019

Purge off this gloom, the soft delicious air,
to heal the scar of these corrosive fires.

Somehow, he made it. Andy Hopper survived the aortic dissection. Ten days after the offensive line coach suffered the critical condition, ten days after being airlifted to Reno's Renown Regional Medical Center, he was allowed to return to his Paradise home.

The team, which had won game three in his absence, received the news at practice and cheered. Hopper vowed to be back on the field soon. Then he got sick again.

On his second day home, he woke up, and his left leg wouldn't move. It turned out, after the aortic dissection, he had suffered four ministrokes. He returned to the hospital, this time Enloe Medical Center in Chico, and his life was back on the brink.

"Just when it seemed like he was getting better, he went from bad to worse," recalled Prinz. "Every day it was a struggle to go to work without him. The kids felt it. I felt it. He's a big man, but the hole he had left in this team was enormous."

The team prayed for him, cheered for him, texted their best wishes to him, and, ultimately, played for him. There was still a season to tackle, and so they did, rampaging all over their fourth opponent, from

Nevada. The Bobcats, in their home uniforms, railroaded the visiting Sparks High School Railroaders, sprinting to a 21–0 lead in the first half before rolling to a 49–0 victory. While Hartley and Harrison were again the offensive stars with 307 rushing yards and 4 touchdowns between them, this game belonged to the unsolvable defense.

That defense was led by a kid they called "Mormon."

"We don't use that term anymore," Josh Alvies would reply.

"Hey, Mormon!" they repeated.

Alvies eventually gave up trying to fight the nickname bestowed on him by his Paradise High teammates, understanding that they meant nothing harmful by it. The middle linebacker was used to enduring friendly ribbing as the rough-edged team's most devout player and one of its two members of the Church of Jesus Christ of Latter-day Saints.

"It comes with the territory," he said. "And I'm proud to be part of that territory."

Josh grew up with his two sisters in a close-knit family whose Paradise roots date back to his great-grandfather's purchase of eighteen acres there in 1924. His father, John, was a Paradise police officer; his mother, Sheris, a senior care assistant. They were deeply entrenched in the community, something that was reflected in Josh's unique involvement with his teammates.

He was their sideline counselor. He was their huddle guru. He was their parking lot psychologist. He always knew the right thing to say. Some of that wisdom he had gleaned at daily religious education classes that required him to wake up at six o'clock in the morning.

"I will just let kids talk, get it off their chest, give whatever advice I could," he explained. "I don't have the answers. but I will be there to listen." Referring to the traumatic events of the past year, he added, "Kids on our team are not themselves. Everybody is hurt in their own ways."

On the field, number 31 was their scrum peacemaker. He would keep things calm. Many times, he would pull teammates from fights, saving them from suspensions.

"He knows how to handle things," said Prinz. "He is our shining light."

Off the field, when hanging with teammates, Alvies gave himself a nickname far less offensive than "Mormon."

"I am the professional D.D.," Alvies said. Designated driver. "Nobody questions me. Most of them listen to me."

It wasn't unusual for him to chauffeur two or three intoxicated teammates home from a party or bonfire, then drive back to load up the car again. Since Josh was a Mormon, alcohol was strictly forbidden, and he would literally extricate himself from the situation if he thought he might be tempted.

"Sometimes it would get to be too much, and I would just leave the party and go to my car and sit by myself for a while," he said.

His one vice? On the field, in the heat of battle, his parents gave him permission to curse, but only at himself. Alvies took advantage of that allowance in a 2018 game against giant Chico High, when he took the wrong angle on a ball carrier, his left knee locked, and he went down in a heap. He was helped to the sideline trainer's table. His parents came down from the stands. Together they cried and prayed. It turned out to be a full tear of his anterior cruciate ligament. He was scheduled to undergo ACL surgery on November 9, but by now you know what happened the day before. Luckily for the Alvies family, the Camp Fire spared their home. Nevertheless, they had to relocate to Chico temporarily, spending the next two months in a home with three other families. Josh slept mostly on the floor.

He finally had the surgery in December and vowed he would play again if Paradise High still had a team. Sure enough, late the following summer, he was back to making tackles.

"As soon as I made one tackle—not the prettiest tackle, but still a tackle—I got over my fears and was back to hitting for my brothers," he recalled happily. "When I'm on the field, it's like a switch goes off, and I know what has to be done." There is no finer example of that than a play in game two, the nail-biter against East Nicolaus High School. Early in the contest, the Spartans beat Alvies badly on what's called a toss sweep play. In the fourth quarter, with the score knotted at 28 apiece, East Nicolaus ran the play again. This time Josh recognized it and was ready. He tackled the ballcarrier hard enough to force a fumble that the Bobcats recovered, and they eventually scored what proved to be the winning TD.

"That's when I knew this team wasn't going to fold," he remembered. "I prayed we would keep doing those little things that would pull us through."

Alvies prayed a lot. Before the season's first game against Williams High, Prinz asked the team who should conduct the pregame prayer. The votes were almost unanimous, and Prinz came to Josh during the final class period on Friday afternoon. Alvies said he would be honored.

"I could see Josh's heart," remembered Prinz. "Everyone could see Josh's heart."

From that moment, it was Alvies who quietly connected the team's spiritual needs and earthly demands in short, soft prayers that focused on both the fire and the football. As the season progressed, he would pray on the field for strength. He would pray on the sideline for safety. Sometimes, when the offense began a drive downfield, he'd excuse himself from the bench, find an empty spot of grass somewhere, and pray for everyone.

"I truly think God used football as a vessel to lift the community," he said. "I wanted to be part of that vessel."

Alvies not only said prayers, but he also helped answer them in a way that was particularly poignant during game three while Andy Hopper was fighting for his life in the hospital. Earlier in the year, the linebacker and his family quietly paved the way for Hopper and his wife, Patrica, to move home.

The Alvieses' modest three-bedroom, two-bath Paradise house survived the fire, but they decided to move to the family's larger property elsewhere in town and put the house up for sale. Meanwhile, Hopper and his family had lost everything and were holed up in a cramped trailer in nearby Durham, desperate to move to a bigger space back in Hopper's ridge roots.

Andy pursued the house through a real estate agent, never speaking to John and Sheris Alvies directly. But it soon became obvious that the Alvies family was going out of its way to ensure that the home was sold to the Hoppers, who offered the asking price of $255,000. Although the Alvieses had been offered more by another party, they immediately accepted Andy's bid. Hopper placed several conditions on the offer; they were agreed to without question. The entire purchase was conducted without negotiation or even a personal conversation. It wasn't until the sale was complete that Hopper learned the truth.

The Alvies family sold it to him not for money, but for love.

"We wanted to know a family's love would still fill the house," said Josh Alvies. "We saw that with Coach Hopper."

Once Hopper and his family moved in, even though they were only casual acquaintances of the Alvieses, the Alvies family continued to watch over them.

When Hopper was first hospitalized in Reno, Josh's mother drove Patrica there to be with him. Then, as Hopper continued to battle in the hospital, his new neighbors were struck by an interesting sight. The kid who used to live there kept showing up in his front yard.

Josh Alvies was praying for his coach while cutting his grass.

"Coach Hopper created a brotherhood among us; he's like a second father to all of us," Alvies said. "When he's hurting, it hurts all of us, and I wanted to do something to help."

The Bobcats were 4-0, and mowing people down.

ELEVEN

The Pain

SEPTEMBER 20, 2019

Awake, arise or be for ever fallen.

After ten days back at Enloe Medical Center for treatment of the ministrokes, three games into a season that was missing him terribly, Andy Hopper finally returned home. But once again, not for long.

Two days later, he was overcome with heat. His heart was filling up with blood. His insides were once again raging against him.

"I don't feel good, honey," he told Patrica. "This might be it."

Soon he was back on another helicopter bound for Reno, only this time a doctor closed the chopper door with a blunt warning.

"I've got to be honest with you: this is bad," the physician told him. "Say your prayers. You might not survive the trip."

Prinz told the players of the latest setback, and he could see it in their eyes. There was heat. There was hurt. They were no longer numbed by Hopper's ongoing medical issues. They were raging mad, at sickness, at loss, at everything, and their next opponent felt it.

The Durham High School Trojans never saw them coming. Literally. With both Hartley and Harrison sidelined by injuries, Prinz gave the ball to 245-pound sophomore lineman Ashton Wagner, number 33, who rumbled for 150 yards and scored 4 touchdowns in a 47–6 pounding.

For the Trojans, the game was out of reach by the time they staggered to the locker room at halftime, down 33–0.

"We're playing with anger. It's the anger of only having one thing to focus on," said Hartley. "It's the only thing that has any importance in our lives. This season is like watching *Friday Night Lights*." He was referring to the popular movie and TV series based on the book by Buzz Bissinger.

Suddenly the supposedly overmatched Bobcats were halfway through their season with a perfect 5-0 record, surprising even Prinz. They had nine new players starting on offense, so they weren't supposed to be this productive. Even with five third-year starters on defense, they weren't supposed to be this stingy.

"I thought we might play evenly with some of these teams. I never expected this," Prinz admitted. "There's a lot I didn't know."

Prinz didn't know that behind those hardened stares, these kids would be playing with such anger. They weren't satisfied with just winning, they wanted to dominate. They couldn't hit hard enough. They couldn't push themselves far enough.

"They're different, they're really different," Prinz said. "They're edgy. They're playing with that anger. They're taking the physical thing seriously. There's so many distractions off the field, they get here, it all comes out."

Prinz didn't know how much his players would delight in the collisions, so violent sometimes that the sound would echo off Om Wraith's burned championship plaque and partially melted scoreboard. He didn't know it would routinely require three opposing tacklers to take down Harrison and Hartley. He didn't know how much the fire would fuel them all.

"We want it more than anyone else who's ever stepped on this football field," Kasten Ortiz said.

They wanted it to the point of sacrificing their bodies for it. Madison

Bergman saw this. She heard this. At times, while taping up a player's injured limb, she even felt it. As an assistant trainer, Bergman lived the Bobcats' pain.

"I would not have believed it if I had not seen it," she said.

Madison, a senior who played on the girls' varsity volleyball team, knew something about resilience. Almost a year earlier, on November 8, she rode down the fiery mountain in the back seat of a cramped Honda, her two best friends in the car with her, their world exploding in front of them.

"Oh my God!" she shouted upon hearing a giant boom. "The KFC just blew up!"

She was not yet an assistant trainer at the time. She was just a sixteen-year-old who was scared out of her mind. She clutched the hands of driver Ceriah Swart and fellow passenger Alaina Hill, who sat in the front. Wild dogs ran past their car window, dancing through the flames. People were abandoning their vehicles and running through the smoke. The three girls had been playing music to distract themselves from the heat, but then the car slowed to a halt in the choking traffic, and the heat was burning at them through the doors, and their cell service died, and so they switched off the music. In a sudden silence punctuated by outside explosions and wails of sirens, they turned to one another, professed their love for one another, and prayed.

It was at this point that Madison Bergman made a vow. She had grown up in Paradise, her family having moved there from Southern California when she was two weeks old. Her mother, Ginger, and her father, Todd, a retired Los Angeles cop, were looking for a better life, and they found it. She felt like her little city was a big family.

"The energy in Paradise is unmatched," Madison said. "The bonds are unmatched."

She wasn't ready to lose that family. She wasn't willing to watch that little city disappear.

So, sitting in that furnace of a car, she made a vow that once they reached the bottom of the hill, she would be part of the city's collective effort to climb back up.

"I wanted to do whatever it took to give our town hope again," she remembered.

She finally made it to Chico, where she reunited with her parents, who had last seen her when they'd dropped her off at school that morning. They had stayed at their Paradise home, which survived the fire, until they knew she was safe. The smoky wait eventually affected her father Todd's lungs so much that he was rushed to the hospital, but he wasn't going to leave his home until he knew his entire family had escaped.

"That's my dad; that was my environment," she said. "We don't leave anyone behind."

Madison decided that she, too, would not leave anyone behind. She kept her vow even as the first postfire months passed, and they moved back into their home with the same depressing realization felt by everyone whose homes had survived. It is a realization mentioned only in whispers, lest someone be considered ungrateful, but it was a truth that resounded through their darkest moments.

It might have been easier if the house had just burned down.

"People who lost their homes say I have no idea," she said. "Well, I have every idea."

They returned to a smoke-damaged, mold-infested structure that needed to be entirely gutted, from the floors to the roof. Clothes, photos, furniture—everything—had to be tossed out, with little insurance money to pay to replace them.

"People think we lost nothing, but we lost everything," she said. "I feel bad telling people I wish my house had burned down. It's terrible, but it's how I feel."

It was a ghost of a house in a wreck of a town, and Paradise no longer felt the least bit like home.

The Paradise city sign when Paradise, California, was still a town of 26,800, before the November 8, 2018, Camp Fire burned the sign and reduced the population by 92 percent.

Satellite image of the November 8, 2018, Camp Fire, which
virtually leveled Paradise, killing eighty-six people and burning nearly
twenty thousand buildings.

Stetson Morgan, running back, hugging mother Emmie after the two escaped the fire in a harrowing drive through a wall of flames.

Morgan holding the face mask remains of two of his helmets that burned in the fire.

Paradise players attend class in a Chico Airport warehouse in the spring of 2019.

Coach Rick Prinz conferring with lineman Jose Velasquez in the combination office–weight room of the Chico Airport warehouse school in the spring of 2019.

From left to right: Coach Bobby Richards, Coach Nino Pinocchio, and Coach Rick Prinz meet around a cot in Prinz's Paradise High office.

The 1947 Ford truck that Rick Prinz pushed down the mountain during the fire to honor a promise made to his father.

The Prinz family football team, from left: Seth, Taylor, Billy, and Jacob.

Angel Lopez, receiver, makes a leaping catch.

Jeff Trinchera, running back, dresses in the parking lot of the Marsh Junior High field used for spring practice.

Tyler Harrison, running back, was bloodied during one of the Bobcats' intense spring practices.

Elijah Gould, lineman, blocks a bag held by kicker Aiden Luna.

Players leaping hurdles at spring practice at the deserted junior high field.

Lining up.

A game of catch at spring practice.

Coach Andy Hopper tells the team the story of the burned championship ring in the Paradise High gym during summer camp.

Lukas Hartley, running back, is overcome with emotion listening to Hopper speak.

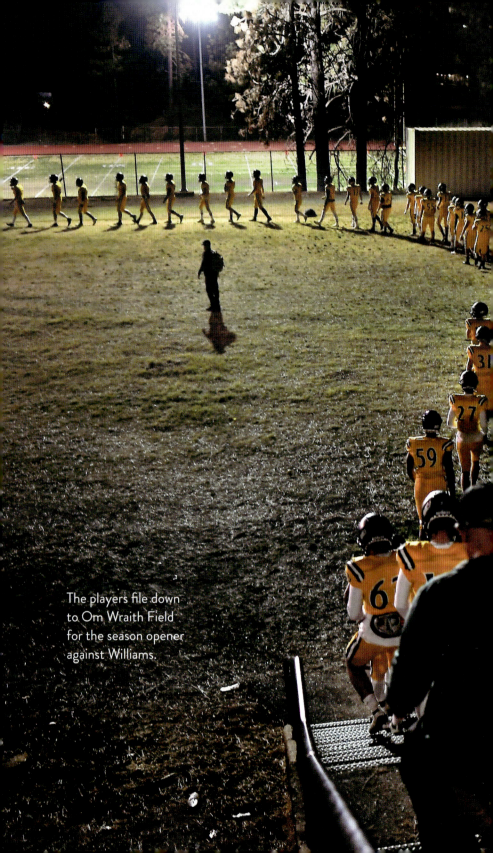

The players file down to Om Wraith Field for the season opener against Williams.

Henry Becker carries the American flag as he leads the team through the stands to the strains of Johnny Cash's "God's Gonna Cut You Down" before the season opener against Williams.

The players bow their heads for the national anthem before the season opener against Williams.

Coach Andy Hopper exhorts kicker Aiden Luna.

Brenden Moon, running back and defensive back, leads the team in a cheer after the opening night 42–0 win.

Brenden Moon despairs in the cold rain after the 2019 Paradise Bobcats' final game, a sectional championship loss to Sutter.

"The second we moved back, I didn't want to be here," she recalled. "It broke my heart, looking out the window, seeing everything gone. It felt like something out of a movie. My town was gone."

But she had made that vow to help bring it back, and, after several weeks of attending class at Paradise High's temporary facility at the Chico Airport, Madison found her opening. It struck her while attending Rick Prinz's sixth-period weight training class, a session that was filled with football players. She heard them scheming about a seemingly futile attempt to regroup and play a season. She saw them practicing in their jeans on the adjacent gravel field. She felt their despair at using begged and borrowed equipment. She watched their expressions as they sadly and continually bemoaned the abrupt end to the previous championship-type season. One of the grumblers was her boyfriend, Lukas Hartley, who, she knew, needed one more chance at running the football.

"I realized, this is where I need to be, this is what I need to do," Bergman remembered. "I needed to be on that field with those players helping bring back the town. I wanted to be there for them and for Paradise."

So, with zero experience, she joined the training staff as one of Chip Schuenemeyer's assistants.

"Chip showed me how to tape an ankle, and I was ready to go," she said.

She may have been ready to go, but she wasn't ready for what she saw. Madison had a front-row seat to the team's pain. The Paradise football culture had always been one of toughness, but in this scarred 2019 season, they were beyond tough. The players had never been so adamant about staying on the field. They had never been so devious about hiding injuries. They had all just walked through a fire. They were not about to be knocked out by a strain or a sprain.

Bergman knew she was working in a war zone when, during one of the first games, Tyler Harrison was flattened. She and others ran out to attend to him.

"I'm fine, I'm fine, I'm fine," he told the training crew as he lay on the grass. But Bergman knew him from school; she knew his eyes, and those eyes were unfocused.

"He couldn't make eye contact with me; I knew something was wrong," she remembered. "But he just jumped up and passed all the tests and kept playing."

Schuenemeyer, a veteran trainer, did his due diligence and properly treated any injury that passed through his sideline trainer's table. But there were lots of aches and pains that the kids never admitted. Offensive lineman Elijah Gould had set the tone by delaying his knee surgery. The rest of the players followed.

"I know it's the culture, I know Chip did everything he could, I know there's nothing more we could have done about it," Bergman said. "But my heart broke for those boys."

Madison Bergman remembered talking to players during the game and needing to remind them of the score. She remembered clapping in their faces to wake them out of a funk. She remembered confiding with players in the parking lot afterward and being stunned that they weren't exactly sure what had just happened out on the field.

It happens in every level of football, even as medical research confirms the damaging long-term effects of concussions. Players everywhere hide their throbbing headaches. Even with strict protocols, they find ways to stay on the field. While football leagues have long done too little too late to promote concussion awareness, those leagues have also been stymied by the athletes they are trying to protect. As Bergman learned while living through one of the dark undercurrents of the Paradise Bobcats' 2019 season, it's hard to save a raging kid from himself.

Then there was the obvious physical pain. Bergman said she saw bruised shoulders, twisted legs, battered feet, dislocated thumbs, yet heard no complaints.

"I didn't understand how they could play," she said, adding, "I still don't."

When she was tending to limping players on the sidelines during games, they would frequently beg her for two things.

"They would say, 'Tape it tighter, Maddy,'" she recalled. "And they would say, 'Don't tell Chip.'"

Bergman felt like she was walking a tightrope between loyalty and duty. But Paradise was built on loyalty and could be resurrected through loyalty, so she refused to rat out her friends.

"It made me so sad, but I knew it would kill these boys to sit on the bench," she said. "So I just taped it tighter."

She once confided to Schuenemeyer about a player's injury, and the team was furious with her. "They said, 'Shut up, we're gonna play, this season is too important, we're not leaving the field.'"

Another time, she was marching over to report a leg injury when the player limped up to her, pleading, "Maddy, don't do this to me." So, she kept quiet, taped tighter, encouraged softly, and did her best to heal what could not be healed.

"When I felt someone shouldn't be out there, they just looked at me, like, 'I've *got* to be out there,'" she said. "I felt helpless. I felt terrible. But football was all these boys had. It was all our town had. I did my best to keep them on the field when they needed to be on the field. I did it for us."

The Bobcats were 5-0, and had no time for comfort.

TWELVE

"I just kept going . . . kept going . . . kept going."

SEPTEMBER 27, 2019

I sung of chaos and eternal night.

Paradise didn't just hit people in games. The Bobcats would go after one another during practice.

This became frighteningly evident in practice the week leading up to game six against the Mount Shasta High School Bears. Feisty Spencer Kiefer and Ashton Wagner challenged each other in a tackling drill, and the much smaller Kiefer got the worst of it, collapsing and losing all feeling in his legs. Prinz called 911. Kiefer was rushed to Enloe Medical Center in Chico, where he regained all sensation and was diagnosed with a pinched nerve in his neck and back.

He didn't miss a play. He returned the next day for the Mount Shasta game after weeping and begging a doctor for clearance.

"I wasn't leaving that hospital until I got that release," Kiefer recalled. "I wasn't going to let anybody take my spot."

It was with this sense of desperation that the Bobcats smoked the Bears. They scored 47 points in the first half alone and didn't need to do anymore in the eventual 47–0 pasting that was born of silence. The three-hour bus trip up to the small city of Mount Shasta was completed with no talking, as per Prinz's long-standing rule. Not only was there no standing on the bus, but also no gabbing, because the coach wanted them to focus. This time, none of the kids complained.

"A mission," said Prinz. "We're on a mission."

As always, the offense revolved around the Bobcats' running game, as Paradise High piled up an amazing 452 yards on 30 carries, for an average gain of 15 yards. Hartley was still sidelined, but Harrison, the breakout star of the season so far, picked up the slack, racking up 148 rushing yards. Stetson Morgan and five-foot-six senior Brenden Moon chipped in 71 and 59 yards, respectively, as the tenacious offensive line busted open huge holes for the running backs to barrel through. Battling alongside tackle Kasten Ortiz was another player with something to prove.

His name was Sam Gieg, and, perhaps more than any other Bobcat, his old life had been changed completely by the fire. He went from stable to wandering. He went from happy to homeless.

"Yeah, it's been a tough year," he said.

It started when Gieg was stuck in a car with his father and grandmother in the middle of a burning city, stuck on the precipice of death.

The junior offensive guard was one of the last people out of Paradise on November 8, 2018. He and his father had to wait for his grandmother. They didn't leave until nearly noon. By then, the top of Skyway Road was surrounded by flames and cluttered with abandoned cars and smoking debris.

"You could feel heat through the window in the car," he recalled. "We were pretty much parked in the middle of the fire, the roads were blocked, we didn't know how we were getting out."

Somehow, his father, Tommy, guided their truck through the mess and ended up in Chico, which is where Sam's odyssey truly began. The fire not only displaced the teenager from his home, but also from his parents. Beginning on the night of the fire—and throughout the 2019 football season—Gieg essentially lived on his own.

"It's the mentality I adopted from Coach Prinz," he said. "No matter what bad things are happening, you keep moving forward."

The first night, instead of cramming into small temporary quarters with his family, he stayed at the Chico house of teammate and fellow junior Caleb Bass, a backup quarterback and outside linebacker. He was one of seven teammates who stayed there, all of them sprawled across the floor on sleeping bags. He thought he would be there for a few days. He wound up living there for six months.

For reasons involving their work and their personal lives, both of his divorced parents decided to move out of the area, with Sam's father relocating around three hours east to Pioneer, California, and his mother moving an hour north to Red Bluff. Gieg didn't want to go to either small city. He wanted to stay home, even though he had no home. He begged them to leave him behind with the stability and security of friends and familiar places. He was born and raised in Paradise. He became part of the football program even though he didn't join until eighth grade. The Bobcats were his family. He couldn't give up on his family.

"I had spent my life with my Bobcat brothers, and I wanted to finish what we started," Sam recalled. "I couldn't leave the team. I couldn't leave the town. I know, there wasn't a town anymore, but I still couldn't leave it. I was determined to stay, even if it meant living on my own."

His parents could have forced Sam to move to one of their new homes, but they knew how much their son had lost, and they couldn't bear to see him lose even more. So, they relented and let him stay. They provided his financial support, promised to remain close and see him often, and, reluctantly, both drove away in separate cars.

"So scary," Gieg acknowledged. "But I knew what I wanted. I wanted Paradise, and I was willing to put up with anything to get it."

For the next year and a half, his physical living conditions were decent. His Paradise family indeed opened its arms. He stayed at Bass's house, then moved in with another teammate, Blake White, then eventually bunked with another buddy.

"I always had people around me who loved me," Sam recalled. "That part made it easy."

It was the actual living that was hard. He had never written a check. He had never used a debit card. He didn't know how to navigate the Department of Motor Vehicles—which didn't matter initially, because for the longest time, he didn't have a car. Money was tight, so he had to get a job while still going to school. Meals were uncertain, so he lived off fast food and the kindness of friends.

And, oh yeah: through all this, he played football for a team trying to resurrect a town.

"I didn't tell anybody what I was going through, I didn't want to make any excuses," Gieg said. "I never missed a practice. I showed up on time. I kept my head down and kept moving. I knew if I slowed down, I would stop all together. I had to stay strong, or else I would fall apart."

During the summer of 2019, he did odd jobs: construction, moving furniture, painting, tending lawns. Many days, he'd work an eight-hour shift and then head straight to three hours of football practice.

"A lot of my kids had it tough," said Prinz, "but maybe nobody had it tougher than Sam Gieg. He was exhausted just trying to live. They all were."

Gieg worked enough that summer to save up $1,000 for a car. But he had never bought a vehicle before and ended up purchasing a BMW with more than two hundred thousand miles on it. It lasted a couple of months until, one day on the way to practice, it began smoking and just stopped running.

"Lesson learned," said Sam. "Being on your own, you learn a lot of things."

When the 2019 season began, he had to take on a second job as a weekend dishwasher at a pizza parlor. So, while his teammates got to relax the Saturday morning after games, Gieg was scrubbing plates and glasses. And while his peers were spending Sunday night preparing for the upcoming school week, he was still at the pizza joint, still scrubbing away.

How did he get his schoolwork done? Sometimes he didn't, and well-meaning friends completed the assignments for him. How did he balance everything? He tried not to think about it.

"I had lost everything," Gieg reflected. "I had lost my normal day-to-day. I couldn't slow down to even think about what that was like. "I was always planning. Where's my next meal? Where am I going to sleep? I was always thinking ahead."

With no safety net, one minor setback could throw him off for the entire day: A forgotten piece of football equipment. A missing home-work paper. He was juggling so many balls, he would often drop one, such as the day he showed up to a football game wearing the wrong-colored helmet.

"I just kept going, kept going, kept going," he said.

Until one night, early in the 2019 season, Sam crashed. Actually, somebody crashed into him. It was in the Bobcats' second game, against East Nicolaus. Gieg, wearing number 53, had just shot up from his stance at right guard when a defensive lineman hit him hard on his left knee.

"Worst pain ever," he remembered.

At first, he played on, though he could barely walk. Sam feared the injury could end his season, and he couldn't handle that. The Paradise trainers asked him how he felt, and, of course, he lied. Only when he was too visibly hobbled to stay in the game did Gieg make his way to the sidelines and crawl onto a trainer's table. Then he did something he had not done since the fire.

Sam Gieg cried.

"I thought, all I had been through, was this how it was going to end?" he recalled. "It couldn't end like this."

Afterward, he was handed a pair of crutches and advised to see a physician. He did not. He could not. He was too afraid of what the doctor would find.

The next morning, Sam showed up to the pizza parlor on those crutches. He worked a shift behind a desk and plotted his comeback.

"Quitting was not an option," he said. "I was going to rest the knee and pray it got better."

Then, during his recovery, Gieg suffered another setback. Inexplicably, he started losing weight—twenty-five pounds in all—and felt tired all the time. One night he woke up in a sweat and phoned his father in a panic.

"Dad, I don't know what to do," he said.

"Drive yourself to the hospital," said Tom Gieg. "I'll meet you there."

So, around midnight, the high school kid drove to the hospital, sat by himself in the emergency room, and was eventually diagnosed with mononucleosis. His father arrived, and they hugged tight.

Then Sam Gieg drove back to the apartment he shared with Blake White. Three weeks later, he was playing football against Mount Shasta.

"Still haven't seen a doctor about that knee," he said. "It gets sore now and then."

He sighed.

"I can handle it."

The Bobcats were 6-0, and handling it.

THIRTEEN

The Smile

OCTOBER 4, 2019

*Abashed the devil stood and felt how awful goodness is and
saw virtue in her shape how lovely; and pined his loss.*

Winning was getting repetitive. It was getting fun.

In the seventh game of the season, the Bobcats devoured the Oroville
High School Tigers in a 57–0 wipeout. It was a night so delightful—
Hartley, back in the starting lineup, and Harrison combined for 372
yards rushing—that Prinz even relaxed his bus rule. On the twenty-mile
trip home to Paradise, for the first time in the coach's career, the team's
cheerleaders were welcomed aboard.

"But the [faculty] advisor was also back there, so it was all good,"
said Prinz.

In some ways, the Paradise cheerleaders faced greater challenges than
the players did. They had to stifle their anger and stomach their personal
losses while putting on smiles for a town that desperately needed smiles.

"What time is it? Game time! What time is it? Game time!"

"All the dogs in the house go woof-woof-woof!"

"Circle up, P-Town, you know, you know; P-Town, the green and
gold!"

Before every Bobcats football game in 2019, Jada Payseno shouted those words as if they were Scripture. The senior cheerleading captain clung to those words as if they were salvation.

"Our job is to show everyone that everything was going to be okay," said Payseno. "In the middle of all these terrible things, we are the smiles."

Even before the fire, Payseno had overcome tragedy. When she was nine, her father, Brad, died of a drug overdose at age thirty-four. The local community had rallied around her family and helped them out of the darkness. Jada had been born and raised in Paradise; her grandparents owned the Jubilee Church. The family could count on their neighbors for whatever support they needed.

Jada's mother, Stephanie, eventually remarried. Four years before the fire, after several moves, Stephanie and Aaron, Jada's stepfather, purchased a house right down the road from the high school: four bedrooms, two baths, trees everywhere.

"We'd found our home," she said.

By then, Payseno had established herself as one of the town's most visible cheerleaders, having cheered for the Junior Bobcats since she was in third grade. If you thought playing football in Paradise was a tough gig to crack, try cheering. For the junior squads, folks would camp out all night outside the Mountain Mike's Pizza restaurant to sign up. Once you reached high school, you had to try out every year; sometimes fifty girls would be vying for just twenty-five spots. It was a big deal. An important deal. And it was the reason that, on the morning of November 8, while running for her life, Jada made certain to grab one thing.

"I wasn't leaving the mountain without my cheer uniform," she remembered.

It all happened so fast. One minute she was happily waking up to the news that school was canceled; the next, her brother was calling to say that something bad was happening.

One minute her mother was saying, "Look, our house is in the middle of town—the whole town would have to burn down for us to be in trouble"; the next, her grandmother was calling with the simple message, "I'm in flames."

The smiles disappeared. The cheerleader got tough. Payseno jumped in her mother's car and ran to gas it up as she was pelted with black ash. Now the tank was full, but so were the surrounding roads. Traffic wasn't moving. Cell service was dead. Payseno was stuck trying to get home. It was ten in the morning, yet so dark she switched on her brights. Then she heard the propane explosions, one after another—*Boom! Boom! Boom!*—and she struggled to keep gripping the steering wheel.

"I was just shaking," she said. "I felt like I was in a horror movie."

She arrived home to find her family waiting to jump in the car. But, first, yes, she ran inside and grabbed that green, gold, and white cheer uniform.

"I knew we had a playoff game the next day," she explained. "I knew I had to be prepared."

Once they began driving, they realized that an entire panicked city had lost its mind. Cars being driven on bike trails. Trucks going the wrong way against traffic. Everyone was crying and screaming.

"I thought, this was it; this was the apocalypse," she said.

Once they reached Chico, the family spent several nomadic weeks there living in a house, then another house, and then a hotel. While their Paradise home had indeed burned, they soon endured another loss when they had to give up their five-month-old Labrador pup named Harley to a pet adoption agency because the hotel didn't allow dogs. Compounding their grief was the discovery that the fire had claimed several beloved teddy bears. The stuffed animals weren't antiques or anything, but in terms of sentimental value, they were priceless, for Jada had dressed them in their late father's clothes.

THEY EVENTUALLY SETTLED IN A townhouse in the middle of a busy area that was the antithesis of their Paradise. It was loud. There was traffic.

There were trains. There was so much missing. Some mornings, Payseno had trouble just getting out of bed.

"Mom," she would cry. "I can't do this."

But she was a cheerleader. She had to do this. She had to bounce up. She had to stretch and smile. She had to cheer not only for the football team but also for the entire town. It was more than some extracurricular activity. It was her calling.

"It was such difficult thing to do, but it had to be done," she said. "We had to put a face on for our town. We had to hype the crowd up. We had to make them believe again."

First, the squad had to believe in itself. Many of the cheerleaders didn't have their uniforms anymore. And because so many kids' families had left town, the three levels of high school cheerleaders were combined into a single thirty-five-girl group. Practices were hectic. Some girls would have to drive more than an hour home afterward. They were stressed. They were exhausted. Girls would be dropped on the ground. Legs would get bruised, elbows scraped. Stress would mess with people's minds. Some game days, Payseno actually forgot her beloved uniform and had to rush back to Chico—with traffic, an hour round-trip—to retrieve it. Those were the times that made her want to stay in bed.

"Some days I would wake up, like, 'Why am I here, and why am I doing this?'" she recalled. "I'm here just to make everybody else happy?"

Still, she kept smiling.

"Smiling and crying at the same time," corrected Payseno.

The captain became a cheerleader for the cheerleaders. She would gather them around and remind them of their mission.

"We're not just here to look pretty," she'd emphasize. "We're here to show people that everything is going to be okay, and that no matter how the town looks, we'll never stop cheering for Paradise."

But, yes, through the smiles, there were tears, before games, during games, after every win. Sometimes those tears were born of relief that the football team was slowly rebuilding the community spirit. Other times, though, they were just tears.

During one particularly rough practice, one cheerleader kept getting a dance move wrong. Finally, she just crumpled to the ground, sobbing.

"I can't do this, I can't go on."

"You *have* to do this," Payseno told her. "This is our therapy. We'll get through this together."

Practice would end, and Payseno would feel like she, too, was on the verge of breaking.

"School is right down the road from my house, yet there was no more house," she said. "I just wanted to go home."

Like most in Paradise in the aftermath of the fire, Jada struggled with anxiety. Just driving in slow traffic brought back burning memories. Yet because she was a cheerleader, she couldn't act out or show it.

"You can't get in trouble, you can't curse, you have dignity to uphold," she said.

Sometimes on Friday nights before games, she would escape the growing crowd and go off by herself to momentarily dream in the quiet.

"I didn't want to talk to anyone," she said. "I would close my eyes and pretend I'm in my room in my old house again."

But then, like Paradise's other brave cheerleaders, she would shake herself out of her funk and paint glittering paw prints on her face, and charge onto the field to energize and entertain.

"We could have all walked out and given up, but Paradise called us back," she said. "We cheer that resilience."

First, before the games, they would bow their heads in prayer, but not the kind of win-the-game prayer you might imagine.

"We would thank God we had this opportunity for one more Friday night," Payseno said.

Then, sometimes, all the cheerleaders would slip a penny in their right shoe for good luck. Not that they needed it. The football season taught them to create their own luck, forge their own path, fight the good fight through both the smiles and the tears.

Then, after the prayer and the traditions, they would roar. Oh, how they would roar, and all of Paradise would roar with them.

"You guys are amazing," a stranger gushed to Payseno during one game. "Thank you for doing this for us. Thank you for making us feel like a town again."

The Bobcats were 7-0, and all the dogs in the house were going "Woof-woof-woof!"

FOURTEEN

Homecoming

OCTOBER 18, 2019

What hath night to do with sleep?

Following a bye week, it was time for the annual homecoming game, and the Bobcats happily gathered in front of another sold-out crowd.

Andy Hopper had survived and was back. Doctors had improved his diet radically, his insides were fixed, and pow, after having missed a month of what was shaping up to be an extraordinary season, he was back on the sidelines, albeit in a lawn chair.

Hopper showed up at practice a couple of days before the homecoming game against the 1-5 Willows High Honkers to bark jokes from the comfort of his chair. He was the same giant presence. The players flocked around him, though stopping just short of hugging his still unsteady body. Instead, they gently touched his shoulder or patted his back, the connection restored.

"I'm so sick of chicken and brown rice, I could strangle someone," Hopper grumbled. "But I see a team that has grown so much since I've been gone. You guys are men!"

The reunion brought Hopper to tears, and he broke into an impromptu speech that held up practice, the kids as attentive and enthralled as ever.

"This season is bigger than us," he said. "There's someone in control over this whole thing. I think the Lord has had his hands in this all along. I'm just happy to be a pawn in that plan."

He was convinced that his medical crisis was triggered by a combination of the fire—which burned everything he owned, none of it insured—and the stress of guiding young men through it.

"I'm a caregiver, and all this I was carrying with me," he said. "I should be dead four times over, but the dissection was God's way of saying, 'This isn't about you. This fire wasn't something they took from you; this was a gift to make you stronger.'"

While Hopper was convalescing, he'd been channeling his energy into going after the team's Facebook critics.

"Hey, I threatened this dude's life on the internet," he told the players. "No one is robbing this from us. Everybody wants to take it. Nobody can take it."

The criticism centered around the Paradise schedule. True, their opponents were mostly smaller schools. But at the time the schedule was constructed, in the spring, the Bobcats had just twenty-two players in jeans and T-shirts. Since then, their ranks had swelled to thirty-nine, with the displaced players returning from other cities, giving them a numbers advantage over most of the teams they were to face.

"When I made the schedule, I had no idea so many kids would be coming back," athletic director Anne Stearns explained.

But she had no regrets. They didn't even know *if* they would be fielding a team in '19, and now people were complaining that their schedule was too easy?

"It's so important that they're winning," Stearns said. "Winning is the one thing that helps these kids feel connected to something. Losing for us would be a hundred times worse than any other team losing. For us to have a chance with the number of players I thought we were going to have, our schedule had to happen (as it did)."

And so, with Andy Hopper back and the Bobcats rolling, the homecoming game wonderfully happened. The stands were filled, the royal court's crowning was grand, and nobody cared that a bank of lights went dark or that the bathrooms still didn't work. This was yet another Paradise football game that really wasn't about football.

"Everybody is talking about the football team. It's the glue that's holding a lot of this together," said Wendy Marsters, a popular, long-time Paradise High biology teacher. "The heart of Paradise is back and beating again."

And beating up on its latest victims. One moment, three Willows High players were limping to the sideline. The next moment, Lukas Hartley was carrying five of them downfield on his back, lumbering forward to gain that precious extra yard. The openings for Tyler Harrison, number 42, were so big, it was as if he were engaged in a solitary Friday-night run. Just one quarter in, Paradise High led, 27–0, and it piled up from there, ending in a lopsided 57–0 score. Every Bobcats benchwarmer, even those who didn't play, spilled onto the field to congratulate this team on a mission.

Once again, Hartley and Harrison roamed the gridiron like monsters in a Japanese sci-fi flick, combining for 248 rushing yards and 5 touchdowns. Amid the celebration, though, there was, as always, a sobering reality, for those two monsters of the backfield were actually two of the squad's most unsettled kids.

Hartley, a senior, lived in a Chico apartment with his older brother, but to save gas, he spent most nights at friends' apartments closer to school. He'd wake up, and he wouldn't be sure where to eat breakfast. He'd finish football practice and not know where he was going to sleep that night.

"Everybody has lost so much, you grab what you have left—which is football—and try to get to the finish line as fast as you can," he said thoughtfully.

The six-foot-one senior lost fifteen pounds at one point because, he explained, "You wake up in somebody else's house, and your big question is, do you eat their breakfast food or do you go buy a nine-dollar burrito?"

Some nights he could barely muster the energy to walk onto the field because "You're wiped out from four hours' sleep on somebody's couch, no food."

Lukas sometimes even had trouble finding the right pregame meal that he would later upchuck nervously on the field. He had to eat to avoid the dry heaves, but he searched for foods that would come out smooth, such as something with rice. The entire process of living made him so quietly furious at the world, it was no wonder that after one game, an opposing team's safety approached Hartley with a confession:

"Dude, after the second quarter, I didn't want to get within ten yards of you because you were coming after me."

Hartley indeed personified the Paradise anger.

"If I was in their shoes, I would think the same thing as that kid," he said later. "Talent, skill, and work show up in the first three quarters. Love takes over in the fourth. We love to hit."

Meanwhile, Tyler Harrison would have loved just to have a regular ride to practice. "Junior" lived in Chico with his grandmother while his displaced parents resided for the time being in San Diego. She owned an old Chevy that always seemed to break down, routinely leaving him at the side of the road, waiting an hour or more for a teammate to pick him up. To save time, Tyler spent many nights sleeping at the home of assistant coach Shannon Magpusao.

"We're playing football like this is all we got, because it really is," Harrison said. "Our team has become so close, like a family. I've never experienced anything like it."

Manuel Rakestraw, the Willows High coach, had a word for it: "Hungry," he said after absorbing the Bobcats' beating. "You can tell they're hungry to play football."

The homecoming victory ended with a more literal example of their hunger.

"Doughnuts! Doughnuts! Doughnuts!" they chanted.

Nino Pinocchio, the defensive assistant, had promised them appropriate treats for every shutout. (Doughnuts, shaped like zeros. Get it?) But although this marked the third goose egg in a row, he still hadn't brought any, because, well, every doughnut place in Paradise had been destroyed by the fire.

The Bobcats were now 8-0, and still famished.

FIFTEEN

The Shirt

OCTOBER 25, 2019

A mind not to be changed by place or time.
The mind is its own place, and in itself.

It was the last touchdown in a flurry of touchdowns. It was Paradise's 58th point of another long night for their opponent, the Red Bluff High School Spartans (yes, another local team with that same name). It happened at the end of the third quarter of another blowout, with almost nobody paying attention.

But when J. D. Webster, number 3, scored on a 10-yard run to give the Bobcats an eventual 59–7 win at Red Bluff, his teammates swarmed around him, pounding his shoulder pads and banging on his helmet. The five-foot-nine spark plug wasn't a star, but he was a symbol. He represented the desperation in their fight and the speed in their step. He was the best example of the urgency they all felt every time they stepped onto the field.

"J.D. is everything that is good about this team," praised Rick Prinz. "A lot of the kids play mad, but he plays furious."

By "furious," he meant that Webster carried the ball only ten times during the 2019 season but gained 178 yards for a team-leading and inhuman 17.8 yards per carry. When he got a chance, he grabbed it—

wrung the life out of it—scoring 3 touchdowns on the ground and catching a TD pass on his only reception of the season.

More than most, Webster was still running from the flames.

"It's been almost a year, and it still feels like the fire was yesterday; every single minute in my head . . . avoiding the flames, thinking I'm going to die," he said, his voice full of emotion. "This helps me escape."

Appropriately, on the morning of November 8, 2018, Webster was essentially saved by his green Paradise Bobcats polo shirt, the one with the yellow logo, the one he wore proudly during the week before games.

The shirt represented all the work he had put into carving out a spot not only as a reserve running back but also as a defensive back. As a sophomore on the junior varsity team in 2017, he led the league in interceptions and was voted defensive MVP. Then he spent the next two years on both sides of the ball, filling the gaps, blazing in and out of the lineup, inspired by something as simple as that shirt.

The shirt reminded him of his idyllic childhood in the Paradise community, where he grew up in a two-story house with three bedrooms, two bathrooms, and a balcony overlooking a backyard that ran into the woods, a garden, swings. The shirt was as comfortable and familiar as the walks he took with teammates to Taco Bell, Dutch Bros. Coffee, the park.

"You can get to anywhere in a few minutes, and everybody knew everybody else," said Webster. "Paradise was really paradise."

J.D. had worn his green Bobcats shirt to school the morning of the Camp Fire. There was smoke in the distance, but he and his father, David, didn't think much about it because there always seemed to be smoke in the distance. Then they pulled up to campus to see everyone running in the opposite direction.

"Why is everybody leaving?" Webster asked a friend.

"The town is on fire!"

His father drove directly to a body shop to retrieve the other family car, which was being repaired. But the shop was locked. So he headed to the nearest gas station, because the needle on his gas gauge was perilously close to E. The line at the pumps appeared to be an hour long.

At this point, a friend called Webster and asked if he had escaped yet. "I'm not even back to my house yet!" he replied with panic in his rising voice. "It might be too late for us."

They arrived back at their house just as his mother, Isabel, was pulling up after evacuating her day care center. The house was surrounded by flames. His mother grabbed a photo book. His sister, Aisa, grabbed the family cat. Webster grabbed his PlayStation. There was no time to save anything else. They weren't sure they could save themselves.

The flames were so close and the smoke was so intense that Webster felt like he was suffocating. He covered his nose and mouth with the only thing he possessed: that green Paradise Bobcats polo shirt.

"Breathe . . . breathe . . . breathe . . ."

He gasped into the cotton as his father's car skidded to a halt in front of a downed powerline. They made a U-turn, but an oncoming driver shouted at them, "You can't go this way, either! This road is on fire. Turn back around!"

Trapped, David Webster spun the car around and drove through yards and flames and veered around that powerline while J.D. kept his faced buried in his Paradise Bobcats polo.

"Breathe . . . breathe . . . breathe . . ."

The heat of the fire made it feel like they were pressed up against a fireplace. Every single house they passed was burning. There was only one car behind them. They were among the last to escape.

Webster finally stuck his head out of his shirt as they were passing the giant sign welcoming visitors to Paradise. It, too, was on fire. But at least they could exhale again, knowing that they would soon be safe.

J.D. pressed the shirt to his face again. This time to dry his tears.

"And that was only the beginning," he remembered.

They stayed at his aunt's home in Chico: six people crammed into a one-bedroom, one-bathroom apartment. For nearly two months, Webster and his sister had to sleep in an empty walk-in closet—Aisa on a mattress, while he curled up next to her on the floor.

"Constant claustrophobia," he remembered.

He spent as much of his days as he could in a nearby gym, lifting weights and doing cardio exercises—anything to avoid that closet. Once he returned to his aunt's he would have to stay up late because his sister stayed up late, both of them lying in that closet reliving their nightmare. Webster's eyes would fill with tears before finally closing. The only thing harder than falling asleep was waking up.

"Every morning I'd wake up thinking I was in my bed at home, memories of all my friends filling my head," he said. "It was terrible, like I had post-traumatic stress disorder. Some people say I did. I think a lot of us did."

He eventually returned to the site of his Paradise home to find everything destroyed. The greenery had turned black. All that was left standing were the trash cans.

"I lived there my whole life, and now there was nothing," he said. "I got angry. I got very, very angry."

The anger continued burning after the family moved into a mold-ridden rental home with owners who eventually tried to evict them during a lull in his father's job with FEMA, the Federal Emergency Management Agency.

"I kept it to myself, but I kept getting angrier," Webster said.

The anger bubbled over several months later when he was attending a spring party at the Chico rental home of teammate Caleb Bass. By then, the displaced Paradise kids felt shunned by their Chico peers, who were upset at the increased traffic and crowded stores. On this night, they felt so shunned they decided to make it a Paradise-only party.

"We decided we weren't going to let any Chico kids inside," Webster recalled.

Sure enough, a bunch of Chico kids showed up, the door was slammed in their faces, and a fight ensued. It was the first sign that the Bobcats were going to attack the 2019 season and lash out at anyone who dared stand in their way. That night, they decided that any further fighting would happen on the field.

"We were going to take our anger out on football," remembered Webster. "It wasn't only going to be about winning. It was going to be about living in Paradise your whole life, watching it burn to the ground, and what were you going to do about it."

It was about a green Paradise Bobcats polo shirt, once bright and fresh, now smoky and gray, a shirt no longer about style, but survival.

The Bobcats were 9-0, and surviving.

SIXTEEN

The Student

NOVEMBER 1, 2019

Our own right hand shall teach us highest deeds.

The Bobcats were feeling . . . bored. They had one more regular season game, against 1-9 Enterprise High, a school in Redding, and, to be honest, they were having a hard time taking it seriously. After all, in their nine games thus far, they'd blanked their opponents five times, while the offense averaged 47 points per game. They were ready for the playoffs to begin. They were ready to finally compete for their dream sectional championship and state title. Little did they know what Rick Prinz knew, that none of that was guaranteed unless they actually finished the regular season unbeaten. Because of their mismatched schedule, one loss could end their playoff hopes.

"That game against Redding, they didn't play worried, but I coached worried," Prinz acknowledged. "We had no margin for error."

When the Enterprise Hornets crept within 21–14 at halftime, Prinz screamed at his team in the dirt lot that served as its locker room. No sooner did the game resume than Tyler Harrison cradled the handoff from quarterback Danny Bettencourt and streaked 57 yards for a touchdown, giving the heavily favored team some breathing room.

Not for long, though. The pesky Hornets came right back with their own TD to close the gap to 7 points once again, and both Prinz and

Hopper could be heard screaming from the sideline. Perhaps their sheer volume refocused the Bobcats because, minutes later, Lukas Hartley dashed into the end zone from 18 yards out, to make it 34–20. Two more touchdowns followed in the fourth quarter, and Paradise High finished its season with an immaculate 10-0 record. Nevertheless, amid all the whopping and cheering, Rick Prinz was . . . concerned.

"No excuse for the game being that close," he told his team after the deceptive 48–26 win. "We need to realize we're walking a tightrope here."

One player certainly knew that. He was the Bobcat who was among the first to jump on his teammates' backs, the first to make his teammates laugh, the team's tiny motor, backup running back Jeff Trinchera. He carried the ball a mere twenty-seven times during the season, scoring just one touchdown, but he appreciated every minute more than most.

"I'm lucky to be here," he said. "I'm going to fight to stay here."

Trinchera possessed fight because he had been in plenty of them. Growing up in Woodland, California, near Sacramento, his fists were flying constantly.

"Fights outside school, fights in the streets, everywhere," said Trinchera. "I didn't start them, but I finished them." Although never large for his age, Jeff always projected a formidable presence.

He blamed much of the fighting on the fact that he, being Caucasian, was the victim of racially motivated bullying from Latinos—one of the reasons that led his father, Jeff Sr., to move the family to Magalia when Jeff was fourteen. Another factor was the same motivation that had lured settlers to the area more than a century and a half earlier. His father, a retired construction worker, thought they could have some fun prospecting for gold.

"We came to a whole different world," said Trinchera. "We left everything in Woodland behind."

Upon joining the junior varsity team, the first difference he noticed was the hitting. Unlike at his high school in Woodland, the Paradise kids hit like grown men, living up to their school motto.

"Other teams I played for were scared to hit, but in Paradise they loved it," said Trinchera. "They were ornery as hell. It was a release. Their way to be nice was to be mean."

But still, he fought. Only this time, like many Paradise players, his adversary was academics. Trinchera could change a transmission and build a staircase, but he struggled with English and math. On one important math test during his sophomore year, he answered only three of twenty-four questions correctly and promptly broke down. He knew. It was the last straw. After nearly flunking out, he transferred to nearby Ridgeview High, a continuation school, to improve his scores. His entire junior year was thus spent catching up on his studies. Since he was no longer attending Paradise High, so couldn't play for the Bobcats.

He would go to Ridgeview during the day, then return home with buddy John Favors to work out at night. In a shed next to his house, they would spend two hours lifting weights. After that, they would box. On Friday nights, Trinchera used to wander down to Om Wraith Field to watch the games from the grass outside one of the end zones. That is, until he couldn't stand watching anymore and left in a hurry.

"It was very hard not to be on that field," he said, sighing. "It killed me. I would do anything to get back to the boys." However, in hindsight, he benefitted enormously from the temporary "demotion," like an overmatched rookie sent down to the minors to work on hitting a major-league curveball. "That year taught me the value of schoolwork," Trinchera said appreciatively. "I wasn't going to mess up again."

FIRST, THOUGH, THERE WAS THE Camp Fire, during which he used his newfound determination to survive. With the flames coming fast, Jeff drove down the mountain in a three-car family caravan with three dogs kenneled in his 1992 Ford truck. They traveled the back roads, so it took seven and a half hours to reach Chico. Several times during that arduous journey, he had to step out of the car amid the gridlock and let the dogs find a place to pee. He stood there holding their leashes and trying to breathe through the blackening sky, shaking the whole time.

"We had just found our place in this great town, and I thought we were going to lose it all," he recalled. "I nearly lost it all."

When Trinchera finally made it to Chico, his truck ran out of gas, and he had to push it to the pump. His family spent most of the next two months living with an aunt an hour away in Sutter before returning to a home that was smoke damaged but spared. Ridgeview High wasn't so fortunate: it burned to the ground. For the rest of the year, classes were held at Chico's Boys & Girls Clubs of the North Valley.

Through all this, Jeff Trinchera had to do his homework.

"Schoolwork is hard for any kid," he said. "But try concentrating on it when everything else is crumbling around you."

After the fire, he improved his grades enough to reenroll in Paradise High and join the team for summer practice, but that was only the beginning of his academic struggles. Like many other Bobcats, he spent the 2019 season trying to survive school.

"I struggled with time management, with what assignments are more important, with just getting the work done," he said. "We were all trying to bring back football at the same time we kept up our grades in school."

After school, there was football practice until six thirty. Then Trinchera would have to go home and chop firewood for ninety minutes. Then he'd walk the dog. Somewhere in there was dinner. Finally, about eight thirty, he would do his homework on a small Dell laptop propped up on a bed in a loft in his house, sometimes staying up until three in the morning.

He called friends for help writing papers. He spent hours poring over PowerPoints. And he sweated his way through keyboard exercises for a technology class. "I had broken so many knuckles in fights, it was hard to type," he explained.

Sometimes Jeff would show up at football practice while still needing extra time to finish an assignment. So, the coaches would send him to the school library in football pants and T-shirt to complete his work. There he might be surrounded by as many as fifteen other Bobcats players. It was not unusual for players to walk off the field and straight to their computers in a school classroom in the hopes of earning a high enough grade to allow them to remain eligible.

As with all displaced Paradise students, the fall of 2019 for the Bobcats football team was an unceasing academic nightmare. It's tough to study when you don't know where you're sleeping. It's difficult to focus on school when the decimated town is begging you to bring back football.

"The toughest thing we did all year was keep these kids in school," remembered Prinz. "The only reason a lot of them even came back was for football. We had to constantly drill into them that school was more important."

Prinz knew school was going to be a problem when he looked at the grade reports in the first semester after the fire. Six of his players had flunked all of their classes. It was the beginning of a constant tug-of-war that lasted throughout the season.

He would bring players into his cramped office and order them to start their homework while he watched. He would march into the weight room, hand over the player's laptop, and tell him to sit on that bench and finish a project. And during practice, besides telling many of them to take their football pads and books to the library, he would demand that some of the more distracted teens do their homework right there on the bench, so he could see them.

"Studying became as much of a team effort as football," Trinchera remembered. "We were tutoring each other at the same time we were making fun of each other for being so dumb."

Over the summer, Prinz set up the failing kids in the Paradise eLearning Academy, an independent study program held on the high school campus and run by a tough but beloved teacher named Christy Voigt. All the kids made the 2.0 grade point average minimum required for athletic eligibility.

Voigt had a daily view of the academic struggles of Paradise students in the wake of the fire. Her standard for success could be found in the way she greeted students every day: "Good morning! I'm glad you showed up at school today!" she would tell them, and she meant it.

The daily triumph wasn't found in passing classes or acing tests. It was found in simply *attending* classes and *taking* the tests.

"I was shocked that they showed up to school every day, because I would think about what they did to get to school every day," she said with admiration. "They're driving in from everywhere, they're not living normal lives, they're scared, they're homeless, their parents don't have jobs."

Like all Paradise educators, Voigt was a hero. She herself had to go through hell just to be there for the students. The Paradise native lost her house in the fire that also razed fifteen other homes belonging to members of her large extended family. She was living with her two young sons across the street from her burned property in a trailer. It had one bedroom. Her boys slept on couches. Yet, like the other heroes, she showed up for work every day fearful of the fate of the city if she did not.

"I firmly believe if Paradise Unified [School District] did not reopen, there would no longer be a town," she said. "We had to be there."

Once she arrived at school, what Voigt confronted on a daily basis was the stuff of both educator challenges and nightmares.

"In education, if a student is not safe or stable, it's difficult to learn," she said. "In the beginning, none of our kids were at that point."

Students would show up and sleep through class because they were living in a car with their mother. Students would ask for second helpings of the free brown-bag lunches—Uncrustable thaw-and-serve sandwich, juice box, piece of fruit—because they were so hungry.

"They had tired, blank faces, they weren't able to learn, they would stare at you, they wouldn't take any information in, most of them were just worried about where they were going to land," she said sympathetically. "I was just trying to comfort them. It was hard for teachers to even talk about grades."

The Paradise football players had the added burden of passing their classes while reuniting the town by performing on Friday nights.

"We have wonderful athletes in all our sports, but Paradise is a football town, and those players felt more pressure because of that," observed Voigt.

When the California Interscholastic Federation (CIF), the state high school governing body, exempted Paradise students from the athletic-eligibility standards because of their unsettled situations, Prinz and his staff continued to work diligently at keeping them in school and on track for graduation. He cared about far more than just the kids' value to the football program; he worried about their futures as young men. He still considered himself a youth pastor and a teacher. He knew they needed the stability of an education. Just because they were eligible to play football didn't make them whole. That's what he preached, and that's how he taught.

"Football was rough, but the fear of not getting the grades was twice as scary," said Trinchera. "All of this while a lot of my teammates were just trying to live."

Some Paradise teachers were uncomfortable with the CIF's waiving the football-eligibility rules, on the grounds that it seemed like preferential treatment. Coach Prinz understood, and, accordingly, he pushed academics even harder, constantly texting his kids with reminders to get to class. But there was only so much he could do.

One time, a teacher contacted Prinz about lineman Sam Gieg's absence from a class during a game day, which meant Gieg would be ineligible for that night's game. It had been just one in a long list of absences by a kid who was essentially living on his own. Prinz angrily summoned Gieg to his office for an explanation.

It turned out that Sam had a good excuse: he had missed that particular class because he had an interview for a job that would help him support himself.

"What am I supposed to say to that?" Prinz remembered thinking.

THE UNDEFEATED BOBCATS LOOKED FORWARD to the postseason for several reasons: During the 2019 regular season, they'd outscored opponents 469–73, a stunning statistic. Also, they had Andy Hopper back and healthy. They had stands full of fans and a rebuilding town full of expectations. Their dream season marched on.

Sort of.

"A dream season?" Hartley said. "A dream season for me would be, at the end of the day, I get to go to my old home."

The Bobcats were 10-0, and yearning.

SEVENTEEN

The Fighters

NOVEMBER 15, 2019

Waved over by that flaming brand; the gate
with dreadful faces thronged and fiery arms.

They say you're resilient. They say you're a fighter. They applaud your strength and idolize your maturity.

But, seriously, when those cheers from the Om Wraith stands end, how do you ever get over having your mother tell you to prepare to die?

That was Paradise running back Stetson Morgan's burden, born of that November 8 morning when he was shoved into the family car as mother Emmie began her blazing drive down Skyway Road toward salvation. The car soon overheated. The flames licked at the windows.

"How many people in this world have to tell your child he was going to burn to death?" she recalled with a shudder.

But that's exactly what she told Stetson as she grabbed his hand across the front seat while they were stuck in a six-hour traffic jam of smoke and ashes and death. They could feel the heat. It looked like the fire was touching the sky. They were literally driving through hell.

"Honey, if we burn to death, I love you so very much," she whispered to her son.

"I love you too, Mommy," Stetson said.

How does a sixteen-year-old kid ever get over that? Even one as tough as Stetson Morgan? The junior running back was only five foot six and 140 pounds, but he had already made his mark as a champion wrestler and as one of the mostly joyously physical guys on the football team.

He was the kid who gleefully flew into the piles. He was the kid who jokingly screamed at opponents. He had a T-shirt that spoke for his essence: "Stick 'em Stets." He loved to stick people, but he did it with a wink and a grin.

The fire stuck him back.

"The fire changed me," he acknowledged somberly. "I'm not myself anymore."

The fire burned everything he owned and landed him and four members of his family in a tiny mobile home on a gravel road in farmlands fifteen miles from Paradise. There was no cable. No internet. On the wall were photos of two dogs and a wrestling medal. Everything else was gone. He wore hand-me-down clothes. He slept on a couch or a chair—that is, when he slept at home. Often he stayed over at friends' houses.

"It's hard," he said. "There's no room."

And there was little money to make things bigger or better. The Morgan family lost its home without renter's insurance. Emmie was a housekeeper who had nine clients before the fire, but eight lost their homes, so she basically lost her job. She had to dig deep to give Stetson $40 for used cleats. He was appreciative. He was also angry.

"He gets angry with me when I don't have enough money, or when I can't give him a ride, or when we have to drive thirty miles to do laundry," said Emmie.

He was distracted. His grades were suffering. He was tired. He was fed up. He was in such need of normalcy that when a stranger donated $1,000 to everyone in Paradise, he used the money to buy his family a coffeemaker and dishes.

"Before all this happened, I saw a bright bubbly kid who just wants to have fun," said his cousin Lily Escobar. "Now he's quiet. He's darker. He's different."

He didn't like to talk about it. His sentences were clipped, though his emotions ran deep.

Stetson summed up the whole situation this way: "This sucks," he said. "This just all really sucks."

In one crunching moment in Paradise's 2019 football season, Stetson Morgan angrily showed his world just how much everything sucked.

At the same time, so did teammate Jose Velasquez, who was fighting through a different kind of hell.

VELASQUEZ, A FIVE-FOOT-EIGHT, TWO-HUNDRED-POUND LINEMAN, was suffering through the loss of not only his home but also his family. Jose, born in Mexico, moved to Northern California as a young child. At age nine, he was taken from his drug-plagued parents and wound up in the Paradise foster home of Tom and Nancy Kelly throughout high school. He always played football. It was his escape. It was his sanctuary.

"It was a place I could get everything out," he said.

When his foster parents' home burned down, he played football. When his biological father was jailed on drug charges, he played football. When his father was released midway through the 2019 season, he thought he might be reunited with his parents and two sisters. But then his father was deported and took his mother and one sister with him to Mexico, leaving Jose and his older sister behind, crushing Jose and making him need football more than ever.

"I did see him a couple of times before he left, and it kind of hurt," Velasquez said of his father. "I gave him a hug, told him good-bye, gave him my farewells, and put everything I had back into the sport."

Leading a defense that allowed only 7.6 points per game—a defense so smart it was calling out other teams' plays in advance—Velasquez, wearing number 54, played like a kid possessed. Which, in fact, he was, swallowed up by the notion of getting even.

"Everybody doubted us, thought we were done, thought we were scattered, thought we were going to be trash," he said. "We have to prove something. We have to make everybody shut up. We have to show them what football was all about."

Together, Jose Velasquez and Stetson Morgan were about to show everyone.

The Bobcats' regular season had ended, but the stress was just beginning. The unbeaten yet not-yet-playoff-bound team was not scheduled to play in the week following its finale win against Enterprise, but it was a bruising week nonetheless.

First, there was the November 8 anniversary of the Camp Fire.

The Bobcats scheduled a film session for that day, the same time they watch film every week. Coach Rick Prinz thought the stability would be good for the kids. Yet more than half the team didn't show up, and the ones who were there, were barely there. While the game film whirred, some kids hid behind sunglasses, some slept, some were on their phones. Then Lukas Hartley walked in late with his dog.

"Not surprised at all that so few people are here," said quarterback Danny Bettencourt. "For many people, it's the anniversary of the hardest day of their entire life. Maybe it's more important just to be with their families right now."

Later that day, Paradise alumnus Scott Galloway hosted a team barbeque on the field, but the players who show up were restless and distracted.

"I've moved on since the fire. I don't even look back anymore. I didn't realize this was the anniversary until my wife told me," said defensive coordinator Paul Orlando. "But our kids definitely look run-down. They're not stable. They're still living on wheels. This just gets harder."

Andy Hopper, who had recovered from his medical crisis and was holding court with a plate of salad on a picnic table, vowed that the day's memories wouldn't be yet another setback.

"It's always been, if something bad can happen to Paradise, it's going to happen," he said. "Not this year. Not now. I just feel like our story is not done yet."

After the barbeque, some Bobcats folks drove to the giant Paradise Alliance Church, where the town held a remembrance ceremony in front of pews packed with somber families huddled together as if attending a wake, which they were.

"It's been a rough year, and many of us are still struggling, still trying to make sense of our new normal," Mayor Jody Jones said to the crowd. "Tonight we gather as a community to turn the page on the most difficult, challenging year most of us will ever experience. I can feel the strength of our community in this room tonight . . . We will do as we have always done. Work hard, make tough decisions, have faith in our ability to withstand the challenges ahead, and once again create our little piece of the world called Paradise."

An elementary school choir sang the stirring Mariah Carey ballad "Hero," and you could hear sobbing and sniffling. The crowd then joined them in singing a tribute song.

"From the ashes we will rise . . . friend and family by our side . . . hand in hand we will survive . . . Paradise strong."

Melissa Schuster, a member of the Paradise Town Council, pulled it all together when she said, "There was more debris removed from our town than from 9/11! Step by step, nail by nail, once again we will make it Paradise."

Step by step, nail by nail, Prinz was doing his part as she spoke. He was not in the audience. He was an hour south, sitting shivering next to Nino Pinocchio up in the cold bleachers at Sutter High School, cheering for the home team to beat the visiting Orland High Trojans.

Why? Because in order for Paradise's team to qualify for the playoffs, the Huskies had to win. Though Paradise went undefeated, it was mathematically possible that the Bobcats could miss the postseason entirely because of the perceived lesser value of their wins against supposedly inferior opponents.

"It's crazy that it's even a question," grumbled Paul Orlando. "How do we not get in? We're ten and oh! Anything less than a state championship will be a disappointment for these kids. These kids want it all. They want to be the best team in Paradise history, and they can be—if they just let us in the playoffs!"

The problem was that, because Paradise had played a makeshift schedule with no league affiliation, it was hard for California Interscholastic Federation Northern Section officials to judge its worth. The Bobcats beat up on smaller teams; what did that mean? Really, how good were they?

The Northern Section football playoffs are divided into five sections based on school size. Paradise, ordinarily in Division 3, played a Division 4 schedule for schools with enrollments between 250 and 499. But because late returnees boosted enrollment to around 525, the Bobcats would have to qualify in Division 3 for schools with enrollments between 500 and 999. This presented an unusual problem for officials. Teams generally qualified for the division in which they played their regular season game. Ranking Paradise against everyone else was comparing apples to oranges. The Bobcats had played the season as outliers. Now they might wind up paying the price. Rankings, to be established at the next day's seeding meeting, were based purely on numbers. There was no emotion involved. Nobody cared about anybody's town burning down. So, yeah, Paradise's coaches were worried.

Prinz was nervous. Pinocchio was nervous. After they returned home from watching Sutter easily dispatch Orland by a score of 33–6, they texted each other until early morning, trying to figure out the formula that would put them in the playoffs. "Nino texted me at two in the morning with the numbers. I woke up and read the text, then he kept texting me like I was his girlfriend," said Prinz.

One text seemed to confirm they were in. The next text wasn't as certain. Pinocchio awakened at five o'clock, filled with dread.

"I thought I had messed up the calculations," he said. "I thought we could be screwed."

Paradise had gotten screwed before. Four times, they had won a sectional championship and yet were not picked to go to the state title. The CIF had since adopted a more straightforward path to a state championship—win three games to win the section, win another playoff game, then compete for the state title—but Paradise High always wondered if the odds were still stacked against it.

"We just don't know," said Prinz. "They couldn't leave us out, could they? We weren't going to be penalized because of the fire, were we?"

On Saturday morning, Prinz and several Bobcats coaches made the hour drive to Red Bluff to show their faces at the seeding meeting. It would be harder to turn them down if officials had to look them in the face, right? In a tiny room filled with other sweating coaches, kicking coach Jeff Marcus stood up and extended one last olive branch to the seeding committee.

"I just wanted to take the opportunity to thank the section for your support and encouragement over the last year. Our community appreciates everything you've done for us," he said graciously.

The committee members smiled, nodded, and then announced that Paradise had made the playoffs.

Barely.

They were assigned 17.38 points, squeaking past Orland High by 0.38 of a point. They were given the fourth seed in Division 3. This meant that the Bobcats would have to beat bigger schools to win the section. They didn't care. They would play the Live Oak High School Lions next week in the first round at Om Wraith Field. They were in.

"Oh my gosh, that was close," Prinz said with satisfaction. "Hard to believe it was that close. I can't believe we had to go unbeaten to get in. I thought we could lose a game or two. The pressure was really on me, wasn't it? If I had known the math, I would have been a lot more nervous. I'm glad they didn't tell me."

Liz Kyle, Northern Section commissioner, emphasized it was not a sympathy vote, insisting, "We followed our policy and process, and these were the results," she said.

Outside in the parking lot on a chilly November morning, Prinz was visibly relieved. Honestly, he was still haunted by his decision to end last season before the playoffs began and was grateful for another chance.

"I sweated it out all week, praying for a chance to make up for last year, scared to death that once again, you give your heart and soul to everything, then have it suddenly end," Prinz said. "I hope people have realized there was no way we could have played last year. They know that, right? If they still don't understand, they don't understand what happened. How could I have asked those kids and coaches to play when they didn't even have places to live? What was I thinking even considering it?"

The playoff berth inspired Prinz to reminisce about his team's first great playoff moment: his first playoff win as a head coach, in 2000, in fourteen-degree weather in the mountain town of Susanville.

Lassen High School's Grizzlies had won twenty-six straight games, and Paradise was regarded so lightly that officials didn't even have a special game ball for them to use during pregame warmups, so they had to borrow an old water-logged pigskin from Lassen.

"We didn't care. We gave the old ball to our quarterback and said, 'Get it done!'" recalled Prinz.

They got it done, all right. Paradise won, 24–21, on Jacob Walton's 2-yard touchdown run with forty-six seconds left.

"Those were the good times," remembered Prinz. "Then there were the bad times."

Although he'd led them to six sectional titles, they'd also lost three sectional championships in the final minutes. The worst was a 26–23 defeat to Enterprise in 2009 when an apparent game-winning blocked field goal attempt was overturned because the celebrating Bobcats had too many men on the field. This allowed Enterprise to replay fourth down, and, three plays later, it scored the winning touchdown.

"We're going to hold our heads up because we're men," Prinz told the *Paradise Post* at the time. "Things don't always go your way in life, and that's the way it is."

Ten years later, he was ready to write a happier ending, as was the entire Paradise High School student body, which, a week later, crammed into the tiny gym several hours before the Live Oak game to inspire its heroes.

"Pep rallies are good because I get out of class," remarked Kasten Ortiz.

The cheerleaders, led by Jada Payseno, were here. The costumed Bobcat, a large furry creature with a permanently opened mouth, was here. The players' names were adorned above the stage in papier-mâché. The lights went off. Students ran in under a canopy of balloons. The eternal high school pep rally chant arose.

"We're from Paradise, couldn't be prouder! If you can't hear us, we yell a little louder!"

The Bobcats players, who wore their jerseys to school, lined up to be introduced. They were not smiling. They were not cheering. They were anxious about the game and increasingly uncomfortable with the burden.

"Pep rallies are all right, they're fun and everything, but I think we can get hyped on our own," said Stetson Morgan. "Wearing this uniform at school puts more pressure on my shoulders to have an impact on the school."

This was the second rally of the day. School spirit was at such a frenzy, they needed two rallies to accommodate everyone who wanted to attend.

"Much louder rallies than in the past," observed Anne Stearns. "People who are here want to be here, not because some teacher ordered them. They are here for a reason. These days, this school takes pride in everything."

Still, his players weren't thrilled, and Prinz was worried.

"This has been a headache all day," he said. "It's so last minute, it's hard. It's their second rally of the day. Why do we have to go to two rallies? My kids have been out of class on their own since third period, and I'm like, 'Oh my God, these kids are going to do something stupid. They don't need this to get fired up.'"

Like it or not, however, the rhyming introductions began, the players trudging across the gym floor wearing baggy jeans and caps askew.

"Meet our outstanding, undefeated varsity football players . . ."

"There's nothing you can do to stop number two, Stetson Morgan! . . ."

"Number fifty-four, Jose Velasquez, he can't help but score . . ."

He can't help but score? Velasquez was a lineman. He looked befuddled. Prinz grabbed the microphone to wrap things up.

"Tonight is a playoff game, it's our last home game, we want you out there in full force. Don't be a slacker! Come to the game!" he shouted to the cheering students.

The pep rally ended, a complete and purposeful and timeless high school celebration with but one thing missing. Nobody introduced Lukas Hartley, one of the team's biggest stars. Where was Hartley?

"He's driving all the way back to Chico," said Prinz with a sigh. "He forgot his pants."

Several hours later, Lukas returned to school, as had his teammates, everyone back in the gym, in uniform, prepared to march down and face Live Oak, a similar-sized school with an 8-2 team from a town of about 8,700.

The players spent this pregame quietly, the only sound their cleats clattering on the gym floor. One player stood apart from the others, fully dressed in uniform but completely alone. It was Sam Gieg, who was still suffering from mononucleosis and thus forbidden to play. But he showed up to play anyway. This game was that big.

"We always felt like this is our sacred place, and, seniors, you may never play here again," Prinz said to his team. "You won't do another walk-through in this gym. What do you feel about that?"

His players shook their heads and murmured. Prinz continued:

"We've been talking this week about playoff mentality, and I want to quote John Heisman, who said, 'It's better to die a small child than to fumble the football.'" Heisman, namesake of the prestigious Heisman Trophy, awarded annually to the most outstanding player in college football, was a legendary college coach from the 1890s through the 1910s. "It's an exaggeration, of course, but I want us to say, 'My job is so important to me that I would rather die than not complete it!' Do your job tonight! Let's finish this! It's been a long haul, a lot going on. Let's finish this! Let's go down and kick some ass!"

The players were quiet no longer. They screamed, rattled their helmets, and pounded their pads. Sensing that not much inspiration was needed, Josh Alvies kept his prayer simple: "Dear Father," he murmured, "let us be safe."

Prinz also wanted to be fast. He decided to begin the game with what he called their Speedwagon offense, named in honor of the station wagon in which he was born. In this scheme, the Bobcats run a play within two seconds of breaking a hurried huddle. The Speedwagon plays were relayed through hand signals by assistant coach Bryson Baker, who was something of an expert with signs.

"Why would I signal in our plays when I can get our varsity baseball coach to do it?" Prinz said with a laugh as he sat in his cluttered office before kickoff.

Standing next to Prinz, serving as a decoy, would be Hopper. Though he wouldn't be signaling anything, he'd be valuable as a giant distraction to opponents who are always trying to steal coaching signs.

"You just play dumb, Hopper," said Prinz, still laughing. "You're real good at that."

Hopper laughed along, as did all the coaches. They were all feeling pretty good about this strategy, even though after a couple of plays, the other team knows what's coming, so it has to be abandoned.

"We'll just be tougher and quicker than they are," Hopper said.

And so they were, storming Live Oak from the start. After the overly amped defense committed three penalties on the Lions' first drive, they eventually held their ground. Paradise took possession, and, well, it was over almost before it started.

On their fourth play from scrimmage, Tyler Harrison motored through the entire Live Oak defense for a 63-yard touchdown run. On their very next offensive play, the sophomore did it again, this time dashing 80 yards all the way into the Lions' end zone.

At halftime, the Bobcats led, 28–0, but Prinz wasn't satisfied. They'd waited a year for this home playoff game, and he didn't want them to blow one minute. He stood among their gasping forms in the dark vacant lot that was their halftime locker room.

"If we go out there and are complacent, it's going to be a long second half," he said. "Seniors, this is your last half on this field. Let's play like it! It's the playoffs! Zero to zero! If you have an ounce of letup, they're going to drive on you! Dig down deep! Forget about the score! We haven't even begun to hit yet, right?"

As the Bobcats bounded onto the field for the second half, the crowd chanted, "CMF! CMF! CMF!"

Prinz was right. His team had not yet begun to hit. For better or worse.

With 9:48 left in the third quarter, in a moment that defined the season, angry Stetson Morgan and angrier Jose Velasquez attacked.

The incident occurred at the end of another Bobcats run. Massive Live Oak defensive tackle Tony Salazar, who stood six foot five and weighed 310 pounds, appeared to twist ball carrier Lukas Hartley's helmeted head, then began punching him. Morgan and Velasquez, who were both in the game at the time, had seen enough. They had heard enough. They had suffered enough. They were going to fight back.

The tiny Morgan ran up and flattened Salazar. Then Velasquez followed Morgan into the pile, jumped on Salazar, and started punching him. Four other Bobcats—Elijah Gould and Caleb Bass, as well as junior wide receiver Angel Lopez and senior tight end Julian Ontiveros—chased after their two teammates in order to hold them back.

The Paradise coaches were furious. By CIF rules, leaving the sidelines for a fight brings an automatic one-game suspension. The coaches knew they were watching six starters—15 percent of their thirty-nine-man roster—vanish in front of their eyes and become ineligible for the next playoff game. The season could be ruined. And the coaches saw this coming. They sensed the darkness in the pep rally. They felt the Camp Fire anniversary anger all week. They anticipated this, yet they couldn't do a damn thing about it.

Hopper waded into the scrum, screaming at his players, "Get back or I'll fight you myself! I knew this would happen! I *knew* this would happen!"

Prinz joined Hopper in the middle of the flailing players and shouted, "Back up! Back up! You wanna fight, take it to the streets! Don't you have any self-control? You belong on the streets if you don't! Get your composure! Have some respect! You're blowing the season!"

The players were now surrounding Prinz on the sidelines, and he had never been madder.

"I'm so sick of this!" the coach shouted, throwing up his hands. "You wanna fight, get out!"

The Paradise coaches, while livid, also knew the truth. The players were also sick of all of this. Yes, they *wanted* to fight. If it meant protecting one of their brothers, they would do so without an ounce of remorse.

Stetson Morgan was the only Bobcat ejected from the game, and the team rallied around him immediately. On the next play after order was restored, a fired-up Harrison scored on a 42-yard run en route to a 56–0 wipeout.

Harrison, who gained 307 yards and scored 5 touchdowns, said afterward, "The holes were so big, I can feel the wind behind me. It's perfect."

His teammates chanted, "Green Machine! Green Machine!" as they waved their helmets high while bouncing up and down on the chilled field.

This was, indeed, the night they became a machine. This was the game that taught them about team. No matter what the cost, Stetson Morgan and Jose Velasquez made the two most important tackles of the season.

"I have my teammates' backs. That's my job, that's who I am," said Morgan. "For the last year, it has felt like the world was against us. All we have are each other. I will not apologize for protecting my brother."

Velasquez was equally unrepentant, explaining, "I've always been a team player, and the other guy was being a dirtbag. He tried to hurt one of my brothers, monkey punching Lukas, and I didn't like what I saw. I knew I would probably get suspended, but I didn't care. I had to get him off. You have to have your brother's back."

Hartley thanked them both, saying, "That guy was choking me, then pounding my helmet, and Stetson came in and literally de-cleated him, right in the ribs. Smallest guy on the field took down the giant!

Then Jose came in and starting pounding. Makes me proud, all the guys taking up for me. The playoffs get really dirty and difficult, a lot of emotions, and these guys are my saviors."

His teammates agreed.

Said Josh Alvies: "We have a fight to take, and we brought it to them."

Kasten Ortiz: "This is a brotherhood. We take that seriously."

The coaches had cooled by the end of the game. They understand that what happened was more important than even the potentially dire consequences. Even though only Morgan was ejected, the other five who came onto the field faced possible suspension after an investigation. Nonetheless, they weren't mad at the Bobcats. They were proud of them.

"It's on now, right?" said Prinz to the team. "It's on now. It's all in."

Assistant coach John Wiggins pretended he was holding up a giant's head as if they'd just slayed Goliath.

"You know what I like about tonight?" he said. "You fought for your brothers tonight. I can't hold that against you. You fight for your brothers."

Nino Pinocchio apologized for once again not bringing an appropriate treat after their sixth shutout in eleven games, saying, "My daughter said, 'Daddy, you've got to buy some more doughnuts.'"

The players howled. The throng of hometown fans, who had made a new tradition of congregating with the team on the field after each game, nudged one another and chuckled. Then Andy Hopper commanded the end zone and everyone quieted down because the big man was becoming emotional. His wife, Patrica, scowling nearby, was still mad at him for risking his health in that brawl. And he was mad at himself for challenging his own players to a fight.

"Let's go end this beautiful story we started," Hopper said, choking up. "Let's end this story we started at a freaking airport on a field of rocks and holes, where nobody thought that we could do anything.

Look what you guys have done with hard work. You have motivated me that I can do anything the rest of my life. You guys have motivated everybody in this town. We're going to finish it right. Next game, look out! You guys ready to go do it?"

"Hell, yeah!" they screamed.

Later, as the exhausted players staggered off the field, some leaning on parents' shoulders, others wrapping their arms around friends, Prinz shook his head and lowered his voice.

"They're just so frustrated with life, they just can't take it anymore," he said insightfully. "They play with an anger that just keeps coming, and coming, and coming."

IT WAS AN ANGER THAT, in one week, would take them sixty-eight miles north to rural Cottonwood for a sectional semifinal game at top-ranked West Valley High.

At first glance, the Bobcats wouldn't have a chance. West Valley matriculated nearly twice as many students, and its Eagles football squad was unbeaten, while outscoring opponents 361–120. Then there was the wild card factor: Would Paradise High lose six players to suspension? If so, the Bobcats would be outmanned by a team with a solid winning tradition. They would be without most of their fervent hometown fans. They would be heavy, heavy underdogs.

"I really want to keep this dream going," Prinz said earnestly. "I'm just not sure how."

The Bobcats were 11-0, and worried.

EIGHTEEN

The Caretaker

Unmoved, unshaken, unseduced, unterrified,
His loyalty he kept, his love, his zeal.

Spic. Beaner. Chink. N-word.

Those were the names given eleven-year-old Shannon Magpusao when he moved to Paradise with his family in 1985. He was the only person of color among his classmates. His family was one of the few minorities on the ridge. The other kids had never seen anything like him before. They weren't sure where he came from. They didn't know what to call him. So they called him everything.

"I was called everything—that is, except the one thing I actually was," remembered Magpusao. "Nobody took the time to realize I was Filipino."

Thirty-four years later, after watching him nurture the most challenged of the Paradise Bobcats through their tumultuous 2019 season—from rescuing kids during the fire to feeding them after games—the townspeople use different words to describe Magpusao.

Cornerstone. Inspiration. Samaritan. Saint.

"He does a lot more than teach the kids football," said Rick Prinz of his receivers coach and equipment manager. "He guides them through life. He's the first one to be there for them in the morning and the last one standing with them at night."

Magpusao is his parents' son: a caretaker at heart, a five-foot-five, unwavering block of granite.

When Magpusao was eleven, his mother, Loly, and his father, George, moved the family from Salinas, down near Monterey, to Paradise to open a nursing facility called the Cedar Glen Care Home. It had twenty beds. The family of five crammed together in one spare room. As hard as it was for Magpusao to deal with the racism in the streets, it was just as difficult for him to handle the workload at home.

He mopped floors, cleaned bathrooms, turned patients over, emptied the bedpans. He saw loneliness. He saw fear. He saw countless people die.

"I learned how to take care of people," he said. "I learned about life."

Magpusao began playing football soon after arriving in Paradise, and that was just as tough, because his father accidently registered him in a division with kids three years older than him.

"Oh, man, did I get hammered," he remembered, laughing at the memory. "But I kept playing."

He was the only minority at his Catholic grade school, and this is where he heard the abuse, from all corners, insults for all races.

"I was literally called every racial slur because nobody seemed to know my actual ethnicity," he explained. "So they called me everything."

This led to fights. Lots of fights. Playground fights. Street fights. His father was a black belt in karate, so he taught his son every possible move, and Shannon tried them all: punching, kicking, winning.

After taking down several bullies, Magpusao kept trying to impress as the new kid, falling in with the wrong crowd in junior high, drinking, shoplifting, dumb stuff. He needed an outlet, and Paradise High sports gave it to him.

"Football and wrestling saved me," he said.

He played nose guard and tackle on Bobcats football teams that weren't very good. He learned that life was about more than winning; it was about enduring. After graduation, he attended several junior colleges and a police academy before moving to nearby Oroville in 1999 to take over one of the several care homes that his parents now owned.

That was the year Prinz began coaching at Paradise. And that same year, Magpusao joined the school as an assistant wrestling coach. The Bobcats fabric fit him. The style of the community fit him. He identified with them.

"I wanted to give back, I wanted to be a part of the culture there," he recalled. "I loved the smaller community feel, the style of living; more rugged kids—folks who always had it a little tougher."

A year after joining the Paradise staff, Shannon was reminded how much tougher when he left for a better and more secure position as an assistant wrestling and football coach at Chico's giant Pleasant Valley High. He lasted three years. He returned to Paradise as the freshman football coach in 2013, joined Prinz's varsity staff a year later, and hasn't left since.

"Chico coaching didn't suit me," Magpusao explained. "The kids in Chico didn't have to work as hard, they had bigger teams, better equipment—every advantage. Plus, they had parents who thought their kid was the best and challenged everything you put them through." In contrast, "Parents in Paradise bought into the football culture. You tell their kid to run into a wall, nobody asks any questions."

Magpusao eventually got out of the care home business, settled into a Paradise house with his wife and four children, and became a fixture at the high school as a coach, campus supervisor, and resident survivor.

"The kids knew he was a caring guy, but a tough guy, and they really listened to him," praised Prinz.

THEY HAD NO CHOICE BUT to listen to him on the morning of November 8, 2018, when he found himself in a crowded and panicked school cafeteria trying to corral hundreds of students in order to shepherd them to safety.

As a campus supervisor, he arrived at school early that morning, and, while he smelled smoke in the air, he didn't realize the extent of the problem until a teacher sprinted past him down a school hallway.

"Where is he going in such a rush?" Magpusao asked another teacher.

"He just found out his house is on fire," came the response.

Seemingly moments later, Shannon found himself literally grabbing students from the cafeteria and shoving them into the cars of parents who had returned to evacuate. Kids were crying. Parents were screaming. It was chaos. Magpusao held firm.

"You get into a zone, you get kids and throw them into cars, kid after kid after kid, you don't even think, you just do it," he remembered.

Busses were still unloading students, who were immediately spun around and guided toward cars. Kids and cars and busses and panic everywhere. When kids couldn't confirm that their parents were coming, Magpusao would shove them into cars of virtual strangers.

"I was like, 'You know this kid, right? You've seen him around? I know this is weird, but take him. Save him,'" said Magpusao.

Then there was the one poor parent whose teenage son could not be found. Teachers combed the empty classrooms and bathrooms frantically. Then Shannon had an idea. Where might a kid have gone to cut class? Where did *he* once go to cut class? He raced down the campus hill to the bike path behind the football field. There, smoking a cigarette with friends, was the lost soul.

"Follow me, now!" screamed Magpusao, directing the boy to the waiting car as they were pelted by floating embers and falling ashes. "Cutting class could have cost you your life!"

At one point, the evacuation was so chaotic that a flatbed truck pulled up, and its driver was surprised when a bunch of kids were herded into the back and told to hang on. When the evacuation was almost complete, Magpusao felt a strong hand on his shoulder. It was another teacher, who knew he had a family.

"Go home, Shannon," he told him. "You've done enough. It's time to take care of yourself."

Magpusao's wife was already down in Chico at her dental office. His oldest daughter lived out of town. So he steered his aging white Dodge minivan to his nearby house, loaded up his three boys and his dog, Champ, and hit the road.

And . . . nothing. Nobody was moving. It took what seemed like hours for him to force his way off a side street and into the traffic jam, where he sat stuck amid the flames, praying for an opening.

His reputation gave him that opening.

A policeman whose kid Magpusao had coached in wrestling recognized him through the smoke-stained driver's window.

"Coach Shannon?" said the cop, pointing to a lane that was just being opened. "Come this way."

Soon his minivan was one of the first cars driving down the wrong side of Skyway, headed for daylight, until a firefighter jumped in his path. It was the father of a kid he had coached in football.

"Coach Shannon, get off this road, it's on fire down there." The fireman directed him to another exit route, this one surrounded by flames but temporarily safe. After being stuck again within a hundred yards of the natural blast furnace, Magpusao finally found enough room to drive to safety. He was filled with terror, thanks, and a realization.

"Twice during my escape, I could have driven into the fire," he recalled. "But both times, I was saved by dads who knew me for my coaching. That can't be a coincidence, can it? That's what Paradise does. That's what it is."

After gathering with his family in Chico, Shannon headed straight to his brother's house in the city of Roseville, near Sacramento, where he, his wife, and his children would spend two months commuting ninety minutes each way to Chico. But they eventually got lucky. In February 2019 they scored a lease on a three-bedroom Chico rental home.

From this gift, Magpusao shared his blessings.

The home immediately became a place for lost or weary Bobcats to crash. Tyler Harrison, who was living with his disabled grandmother after his parents moved to San Diego, basically moved in.

"Staying with Coach and his family, it's such a safe space," said Harrison. "I think all of us were looking for a safe space."

If Magpusao's house was a hotel, then his white Dodge minivan was a school bus.

Once school began again in Paradise in the fall of 2019, Magpusao would leave the house forty minutes early to pick up at least three other displaced kids, sometimes more, and drive them to campus. He would often bring a breakfast of eggs and bacon and sausage, and the hungry kids would devour it. Or he would bring everyone a breakfast burrito.

"I have this weird rule: if I'm eating, you're eating," he said. "And some of these kids really needed to eat."

After a long day of school and practice, Magpusao would load up the van again, only this time there might be ten people packed inside—so many uprooted and distressed and frazzled kids needing a ride home. He didn't leave campus until no more kids were waiting. He rarely completed the drive without stopping at someplace like Taco Bell for a communal dinner.

"It was these kids, man. I just couldn't leave any of them; not after all they've been through," he said. "And if there's one soul left on campus, and he insists on waiting for someone to pick him up, I wait. You'll never feel right if you haven't taken care of that last kid."

One time, Shannon actually squeezed a dozen kids into the minivan, with a specific order if they passed any police:

"Everyone not in a seat, duck," he told them. "Everyone in a seat, sit tall!"

Magpusao suspected that most of the kids were suffering from post-traumatic stress disorder. He noticed one boy in particular who always seemed to be standing alone in the parking lot long after football practice ended before grabbing a different ride every day. One evening Magpusao sat him down for a chat.

"Where are you going every night?" he asked him.

"Depends on the night," the boy said.

Magpusao learned that the player's parents had moved away, yet, like many, the teenager was reluctant to leave the warmth and stability of his football team.

"This was a kid bouncing around, a kid floating around, sleeping on floors, sleeping on couches, all because he wanted to be part of us," recalled Magpusao. "This was a kid who had a need to show up somewhere, with somebody, every day. This kid was really hurting."

Magpusao listened, learned, comforted, then told him what he told all of them:

"If you ever need a place to go, just jump in the van."

NINETEEN

Moonshot

NOVEMBER 22, 2019

*Some natural tears they dropped, but wiped
them soon; the world was all before them.*

Brenden Moon knew there had to be a light. Somewhere, there had to be a light. Even a tough kid like him needed a light. Even a five-foot-six hitting machine needed a break.

"So dark," said the Paradise backup running back and safety early in the 2019 season. "When is everything going to stop being so dark?"

After a year of unimaginable sorrow, hope finally flickered for Moon four days before the Bobcats' sectional semifinal game against West Valley High. All six starters involved in the Live Oak brawl had indeed been suspended, and everyone on the field was suddenly looking at him. The Bobcats' defense would be decimated by the absences, and he would have to start at safety against the powerful Eagles' passing game. He had never started in a playoff game. He had never been asked to be a star. Yet it was suddenly on him.

The night before the game, Caleb Bass texted Moon with a simple command: Get this shit.

Moon gulped. This could be his moment. This could be his light.

His darkness, as with all of their nightmares, began a year earlier, when he was riding to school with teammate Dylan Blood in Blood's eighteen-year-old truck. They suddenly noticed that it was raining fire.

"Embers were falling from the sky, everyone was running out of school, freaking out, shouting that their houses were burning down," Moon remembered.

There was no time to return home and accompany his foster parents of nine years, Debbie and David Turnbow, down the mountain. So he and Blood would have to do it alone. And possibly without gas.

"There's a line of five cars at the gas station," he remembered. "When we get to the front, they turn the fuel pump off. Dylan runs outside cussing, his gas gauge didn't work, we had no idea whether we would make it. We were ready to get out and run."

The truck eventually made it, but the horror of the five-hour drive will remain with Moon forever.

"Miles and miles of cars, people crying and trying to hitch rides, the whole place looking like a nuclear bomb went off. You could feel the vibration from the heat. We thought we were going to die," he said.

His darkness continued when he reached the bottom of the hill and learned that his house had partially burned. He didn't want to hotel-hop with his foster parents, so he spent the next several weeks sleeping on air mattresses and sharing beds with teammates. It was several months before his foster family could return home. During that time, Brenden thought he was living through the worst of it. He was wrong.

"With all the stuff that happened in my life, I like to power forward, aim toward the future," he said. "The fire was just another obstacle. But things just kept happening, and you wonder, 'When is it going to end?'"

MOON HAS WONDERED THAT SINCE he was taken out of his home and placed in the foster care system at age nine. His mother, Sarah Lytle-Moon, suffered from drug addiction and bipolar disorder, while his father, Joey Moon, developed an addiction to crystal meth. Brenden would have only occasional contact with them in the ensuing years, but he was particularly close to his mother.

"She was a very good woman with a very bad problem," said Molly Holmes, Moon's grandmother, who lived up the road in Magalia with Moon's father.

Brenden can still hear his mother's voice and feel her embrace when she rocked him as a child.

"My mom sang me a lullaby every night, every single night," he remembered. "I ask other kids, and nobody's mom did it every night. That is the one time I felt completely safe and centered."

Though he would find stability in the Turnbows' home, Brenden had previously struggled in foster care due to behavioral issues.

"I've had chairs thrown at me—and I've thrown chairs," he acknowledged. "I used to be labeled one of '*those* children.'"

His feelings of insecurity manifested themselves on the football field, where Moon never held back.

"I play angry. Always," he said. "I'm not nice on the field. Ever. I really, truly let everything go. All my anger and worries go away, and I just play the game."

Hopper said the team understood Moon's motivation even as it braced for his hits during practice.

"He has a chip on his shoulder that is really big," said Hopper. "He acts like he doesn't care about anything, but he is one of the guys who cares the most."

Through it all, Moon maintained a soft spot for his mother, even as she was in and out of rehab and jail.

"She loved music, loved Disney, loved to sing, loves animals, loves cats," he said. "She always told me she's going back into rehab, going to get better."

Entering the 2019 season, the last time he had seen her was after a game the previous year. The last time he had talked to her was several months earlier. Sarah always seemed to call during practice or workouts, and he was never near his phone.

"When you get a call from jail, they don't allow you to call back," he said. "I hate that. I feel terrible about that."

He sometimes felt bad that he didn't visit her in prison, but what could he do? He was a high school kid with no car. Really, what could he do? Sometime his darkness seemed so unfair.

"I had no time to visit her, because part of me didn't want to deal with all that," he said frankly. "But part of me was, 'That's my mom; I should be there for her!' It's a constant struggle."

Hours after the Bobcats' September 6 win against Gridley High, Moon was awakened by a call from his half sister, Delilah Ogarrio.

"She came right out and said, 'It's about our mom; our mom tried to kill herself,'" he recalled. "It was like a bad dream."

Sarah Lytle-Moon had tried to hang herself in the Butte County Jail. Authorities rushed her to the hospital, where she was placed on life support. Brenden and Delilah hastened to her side. Brenden remembered the tubes coming out of her body, the handcuffs fastening her to the bed, the swelling around her eye from a jail fight. But mostly he remembered the burn marks on her neck.

"As soon as I walked in, it was, like, crazy to see that," said Moon. "I've never seen a human being try to hang herself. I've never seen a bruise like that. I was wondering, couldn't they have at least covered up her neck?"

It was a lot for a seventeen-year-old kid. Moon embraced his mom through the handcuffs and stayed by her side for two days.

"It's amazing the strength he found at such a young age," marveled Ogarrio. "He's taken everything that life has thrown at him and made the best out of it."

On September 9 Sarah Lytle-Moon died at age forty-three.

Brenden's sobbing last words to her were a promise.

"She knows I loved her, and I know she loved me," he said. "Now I play football for her. I play life for her. I will push through for her."

He says that finding the strength to persevere would have been all but impossible without his Bobcats family. The brotherhood forged by fire surrounded him during his personal tragedy.

"Football got me through this," he said. "These boys are my brothers."

The first person he told was his teammate Blood, and, after he described his mother's bedside scene, they cried together.

"Nobody should ever have to see something like that," said Blood. "I told him, 'You are one tough kid.'"

Brenden received a ride to his mother's memorial service from Lukas Hartley, who simply left school during lunch period to be there for his heartbroken teammate and friend.

"Brenden Moon has gone through hell this year," said Hartley. "Like all of us, this team is his temple, and this season is his savior."

Moon returned to football practice just three days after his mother died, living proof of his pledge to her. His teammates surrounded him with hugs. A crusty coach put his arm around him with an offer.

"If you need a break, anything, just ask," Nino Pinocchio told him.

"Coach, I want to be here," Moon replied. "I want to be *here*."

Prinz then approached him with another hug and a question about how he was doing.

"Coach, I just want to practice."

The next night against Sparks High School, the fourth game of the season, Moon cried through the national anthem. Then, at the end of an eventual 49–9 rout, Coach Prinz knew what he had to do: give Brenden Moon a chance to score. On the Railroaders' 19-yard line, quarterback Danny Bettencourt handed him the ball. Moon scurried for 18 yards. They gave him the ball again. Touchdown. Tears everywhere.

"We were going to spend the rest of the game giving him the ball if necessary," said Prinz.

After the final whistle, Moon hugged Prinz with a quiet "Thank you."
Prinz hugged him harder and said, "No, thank *you*."

Still, the darkness remained. Moon loved his teammates, but he didn't want a pity party. He respected his coaches, but he didn't want anything to be handed to him. He wanted to earn it, to somehow carry this team just like it had long carried him. He wanted a chance to create his own light and run to it. But how could this happen? How could a benchwarmer ever shoulder such a burden?

Suddenly, days before the biggest game of the season, circumstances pushed the benchwarmer to the fore. Could Brenden Moon really have a moment? It seemed too improbable to be real.

MEANWHILE, PARADISE HAD SPENT A tumultuous week trying to keep those starters from being suspended. Anne Stearns, the athletic director, made her case to four local school and CIF administrators. The meeting, held in a Pleasant Valley High classroom, was delayed because they couldn't get the internet to work. None of it mattered, though. The rules were clear, sealing the six players' fates.

"Their line was, 'you shall not leave the bench, period,'" said Stearns. "Our line was, look at the intent. Our kids were just protecting each other. All they believe they have is each other. They've been through hell. We've admired them for being brothers, now we're penalizing them for acting like brothers. It makes no sense."

The announcement was made on Tuesday, four days before the West Valley game. Starters Stetson Morgan, Jose Velasquez, Caleb Bass, Elijah Gould, Angel Lopez, and Julian Ontiveros would miss the most important game of the season. A running back, receiver, and guard on offense. Two starting linebacker/safeties and a lineman on defense. A half dozen holes against the top-ranked team in the division.

"I'm sad, I'm just sad, for everything these kids have been through," said the superintendent of the Paradise Unified School District. "It saddens me that it comes to this. I'm having a very hard time with it."

Prinz stood on a darkened field after their Tuesday practice and shook his head.

"We're in shock," he said. "Our list of obstacles just goes on and on. It's been such a tough year. The kids have consistently tried to hold this team together and be there for each other, and now they're getting punished for it."

His players were looking tired. They were sounding fed up. They were feeling sad. Could their season really end because of the very thing that had held it together? The irony was not lost on them. On Wednesday they trudged to practice with their heads down.

"Last year we all had to run away together from a fire, and everybody loved us, but now we all ran together into a fire, and they want to hurt us," said Velasquez. "They're gonna pull that shit? What is that?"

Those crazy mountain folk had perceived the stares and snubs from the valley for so long, they were starting to believe them.

"Why does everyone hate Paradise? Why do they want to stop our amazing season?" asked Gould. "This is why: everyone this year is a little angrier, they want to fight a little harder. We're past the rage point."

The suspended players were banned from the sideline. They all vowed to be standing on the sideline anyway.

"If I have to, I'll steal a car and get my ass up there," said Velasquez.

"I'm going, because I don't want to miss watching our backups whip their ass," said Gould.

Meanwhile, the coaches were bemoaning another issue. Even when people tried to be nice to them, it no longer felt that nice. They learned that a commemorative coin would be used for the coin toss before the game. On the side imprinted with a Bobcat, it would read: "11/8."

"I guess we're just getting tired of being remembered only for the fire," said Andy Hopper. "This season is about creating something different."

During the week of practice, that creation was loud and violent. The hits had never been harder. The screams and taunts among themselves had never been stronger.

On Thursday night Coach Pinocchio gathered the emotion-wracked team at his feet and fanned the flames with a message.

"Those of you who can't play tomorrow, stand up!" he said, and the six players stood. "How much do you love these guys? Would you not do anything in the world to get them another game, to give them the opportunity to redeem themselves? Could you not do that? I would."

The players were murmuring now, and Pinocchio continued.

"Honest to God, I can't think of very many things that would motivate this team more than this right here: standing up for these guys who have fought with you all freaking year, bled with you, been through the same fires, the same damn things. You have it in your ability to give them another opportunity.

"When you are fighting for each other, with nothing else in the world mattering, there's no power better than that. And that's what we're going into tomorrow to do!"

The players were cheering now, banging the shoulders of the suspended six, ready to answer Hopper's screamed questions in a sequence that epitomized their lives.

"You guys pissed off before this game?"

"Yes, Coach!"

"You going to be pissed off during this game?"

"Yes, Coach!"

"You're going to be pissed off after this game even if we win?"

"Yes, Coach!"

"Admit it, we're living pissed off right now! Let's go kick their ass— that's how we get it out!"

Next, Gould stood up and addressed the team: "Why does everyone hate Paradise?" he asked rhetorically. "Because we just hit people!"

Hopper chimed in with another observation that cut to the heart of the belief system that had fostered a culture.

"Look down at a lot of those valley boys, all the toys they have, all the things they have down in the valley. You know who they want to be like? Us!" he said. "They want what we have, because what we have, you can't take away. What we have is inside of us; it's not anything you can own or buy. They hate us because they want to be us! Does that make sense, boys?"

"Yes, Coach!"

Later that evening, they were briefly locked out of the cafeteria for their weekly Thursday night meal because somebody had stolen the keys. Yeah, it's never easy being Paradise.

A DAY LATER, THE TEAM returned to the cafeteria for the afternoon pre-game meal arranged by Greg Kiefer. The menu? Lunch meat sandwiches for strength, bananas to avoid cramps, pickles for fun. The players devoured it as if it were their first meal of the day. For some, it was. They were still griping about the suspensions. They were still vowing to get even.

"When you've been through hell, this is just a speed bump," said the elder Kiefer. "Don't forget, they're not just playing for themselves, they're playing for the whole town. There's a lot of power behind them."

Midway through the snack, assistant Bryson Baker approached Prinz with some potentially bad tidings.

"I've some news about some of our players' grades," Baker said.

"Just tell me if they're all playing tonight," Prinz half pleaded.

"Yeah, they're all playing."

"Then for now, I'm good," Prinz said.

He'd deal later with the obstacles facing all Paradise High School students as they attempted to study in cramped living spaces, amid distracted parents, while driving an hour each way to school. For the football players, who had to contend with daily practice sessions as well, the obstacles were higher still. Prinz knew that the kids on his team had grade problems. Straight As? Forget it. He was just thankful that all of them were eligible to play Friday night.

The coach retreated briefly to his office to scrounge through his desk for those commitment cards everyone filled out in August at the final team practice before the season opener. He found them. Tonight he would use them. Tonight he would use everything.

"So many distractions," he said. "So tough on everybody."

As the players boarded the bus for the drive to West Valley's Cottonwood Stadium, they passed a purple paper banner hanging from one of the school walls. It had been placed there to inspire the student body throughout this unsettled year. But by now, it was tattered, frayed, and nearly falling down, though its message was still legible: "If life gives you lemons, make lemonade."

As the bus cruised north, darkness approached quickly, the weather turned cold, and the landscape melted into black nothingness. They drove past almond farms, dairy farms, wheat fields, feed stores. Close to their destination, along a dusty farm road, players pointed out the windows at three makeshift cardboard signs flapping in the wind:

"Go Eagles, Beat Paradise!"
"Go Eagles, Go Fight Win!"
"Pasture Pride!"

Suddenly, seemingly from out of nowhere, a collection of school buildings appeared. Tailgating festivities were already under way in the parking lot. The stands were already half full.

The Bobcats were ushered into a cramped girls' locker room. Prinz eyed the place suspiciously.

"Don't leave anything in here," he said. "Get dressed and put your stuff in a bag and take it outside."

One player brought the wrong stuff. Sam Gieg had accidentally packed his yellow helmet instead of the prescribed green one. Hopper threw up his hands.

"The one player I'm worried about messing up, he just messed up one more time," Andy said with irritation, but he knew of Gieg's incredible daily stress, so he shrugged, smiled, and found the kid the proper helmet.

They went outside to begin pregame drills on a nearby soccer field. But moments later, the Eagles marched out to the same space, forcing Paradise to move to the other end of the field.

"It never fails, they always try to get in our heads," said Hopper. A wicked smile crossed his face. "We don't get in their heads—we just hit 'em."

Amid the sounds of cleats skittering across the grass, there was the noticeable noise of retching. Lukas Hartley was vomiting. Tyler Harrison, too. And J. D. Webster. The entire backfield was vomiting.

Prinz waited for their stomachs to settle, then gathered the players around him in a circle in the dark.

"Everybody thinks you're down, there's people dancing on our freaking graves!" he said. "You guys that are in there tonight, just frigging go for it."

Then he stared directly at a tiny benchwarmer who'd just become a starter through attrition.

"Moon, go for it, dude, right?"

There it was. The first acknowledgment that the team was relying greatly on bedraggled backup safety Brenden Moon. He looked up and nodded. He was actually smiling. It had been a nightmare year, and he was ready to wreak hell on whomever.

Prinz reached into his pocket and pulled out a few wrinkled commitment cards. They'd been written by the suspended players.

"They want to win a championship—that's what they all wrote!" he said. "They can't do that if you don't win tonight. Their freakin' attitude is Paradise, right there. That's what we have to play with tonight: reckless abandonment. Just freaking go for it and let nothing hold you back. Can we do that?"

"Yes, Coach!"

Josh Alvies offered the usual quiet prayer, then Nino Pinocchio sent them to the field with a pregame speech as momentous as the moment.

"Three letters, sumbitches! Three fucking letters! It has one meaning, one heart!" he shouted. "The freaking letters are tattooed on my body, and I hope half of you will probably do it yourself someday!"

Pinocchio continued, his voice crescendoing: "Let me tell you about their meaning. It doesn't matter if they make us come up here in the middle of a dumb, fucking Podunk town. It doesn't matter if they fucking take six of our players away from us. It doesn't matter if they freaking burn our freaking town down—it just doesn't matter. When we put those green helmets on, and we take that field, they will find out what those three letters are all about."

His eyes were filling with tears as he finished with a shout: "CMF! You got to love it, you got to feel it, it's got to be in your soul! Bring it up here! Let's go!"

The players jostled against one another as they hurriedly fell in line to march onto a yellowed field with crooked goalposts and uneven yard lines. The stands were full, the surrounding fence was lined with people. The atmosphere was perfect.

And so, quite unbelievably, was the Paradise football team.

MIDWAY THROUGH THE FIRST QUARTER, the offense drew the Eagles defense offside on a fourth-and-1, costing West Valley a 5-yard penalty and a replay of the down. On the very next play from scrimmage, it would cost the Bobcats' opponents much more than that, as the young six-foot-three phenom Tyler Harrison scored on a 47-yard run. An expletive-laden cheer went up on the sideline. It was Jose Velasquez. He and the other suspended players had showed up, as promised.

The defense held West Valley on fourth-and-7 at Paradise High's 17-yard line. Then something incredible happened offensewise: Danny Bettencourt hit Ben Weldon on a 21-yard toss to keep a drive alive. The Bobcats almost couldn't believe their eyes.

"We passed!" they screamed from the sidelines. "We actually passed!" Their excitement was understandable: Rick Prinz's offense rarely threw the ball more than a half dozen times per game, and there had been several contests this season in which Bettencourt got to cock his arm and send the ball airborne just *once*.

As the game progressed, the teams exchanged possessions, with the makeshift Bobcats defense stopping the Eagles on every big play. Then—who else?—Harrison broke free for a 73-yard touchdown run to give Paradise High a 14–0 lead at halftime in a game that was far more hard fought than the score would suggest.

Something big had to happen for Paradise to clinch this. The half-time speeches predicted that moment was upon them.

Shouted defensive coordinator Paul Orlando: "The most important drive of the second half is the first drive! You will take their will if you shut them down!"

Hollered Pinocchio: "Make them make a mistake! Whoever blinks first is going to lose!"

West Valley blinked first, stumbling into the wide eyes of—yes, indeed—Brenden Moon.

After converting on a fake punt and piercing deep into Bobcats territory, Eagles quarterback Kitt McCloughan lofted a slant pass to teammate Dylan "Charlie" Booth with 7:35 remaining in the third quarter. It was a well-designed play that should have gone for big yardage—maybe even a touchdown—and help them seize the momentum.

Except number 8, Brenden Moon, got smack in the middle of it.

"I figured they were going to pass," he recalled later. "I did my read, went to my spot, the quarterback ended up throwing the ball straight to me—completely stunned me, made eye contact with me, looked straight at me, threw the ball straight at me!" It was as if he still couldn't believe it.

Moon caught it. Moon picked it. Moon owned it. Intercepting the errant pass at the Paradise 25-yard line, he raced 75 yards through the chilly fog to score the game-clinching touchdown without a single Eagle laying so much as a finger on him.

"I was like, 'No way! Did he just throw me the ball?'" Moon remembered thinking. "Then I'm like, 'I'm going to score a touchdown, right? That's crazy!'"

"It was pretty sick," said Hartley. "He didn't even break stride. Looked like he'd been doing it for years. It was just perfect."

Halfway to the Eagles' end zone, Moon glanced toward the visitors' sideline, where his overjoyed teammates were sprinting along with him while screaming his name: "Moon! Moon! Moon!" They were the ones who'd helped him navigate through the darkness; now he was paying back his brothers.

Finally, Brenden Moon was basking in the light.

As he crossed the goal line, the crowd got into the act. Hundreds of Paradise residents filled the bleachers and leaned against the fences, including his grandmother and his father, who never missed a game. "Moon! Moon! Moon!"

Brenden, so caught up in the moment, never heard them, however. When told later of all the cheering from the stands, he was taken aback. How was it that so many people knew who he was? Hadn't Paradise been decimated by the fire one year ago? Why were so many people from his hometown even here?

"Really, they chanted my name?" he asked. "Oh, my. Really? Oh my goodness."

As Moon stood triumphantly in the end zone, he thought of one person in particular. He patted his heart and pointed at the sky. Then he heard another chant.

"That's for your mama!" his teammates shouted. "That's for your mama!"

Yes, it was for his mother. How did they know? Moon started crying. They all started crying.

"When I crossed the goal line, automatically my mind went to my mom, but I didn't think anyone realized it. I thought everyone had forgot. It felt good to know that they care," he said gratefully. "It felt really good."

They cared about Moon. They cared about one another. Rising from the ashes of their homes and the rubble of their roster, the once-shattered Bobcats had grown to care about little else other than this family unit that had made them so strong. "Oh my goodness," indeed.

"After all we've been through, we have each other's back in everything; we play for each other," said Blood. "It's win or death. If it wasn't for Paradise, none of would be who we are. Paradise is something within each of us, and we're fighting for it together."

Moon's heroics gave the Bobcats a 21–0 lead that eventually became a 28–13 victory. The scoreboard glowed through the mist, their biggest upset in many years was complete, and the team and town once again rushed the field together.

"Green Machine! Green Machine! Green Machine!"

"CMF! CMF! CMF!"

And, of course: "Moon Pie! Moon Pie! Where is Moon Pie!" Brenden Moon disappeared inside one giant hug after another.

The Paradise High Bobcats were now one win away from the holy grail: the sectional championship, with the title game to be played the following Saturday against the powerful and higher-seeded 11-1 team from Sutter.

"We're going dog hunting!" shouted Orlando about the Huskies. "We're going to show Sutter what Paradise is all about!"

They believed it now. It was close enough to touch. If they could beat West Valley with a depleted roster, they could beat anybody.

"I was nervous we lost the six players, but we made a statement to our section, to our team, to our town: we're coming!" exulted Harrison, who gained 243 yards on just 19 carries, averaging nearly 13 yards every time he touched the ball.

A famous alumnus was impressed.

"They're pretty fucking good—it's unbelievable," said Jeff Maehl, the retired NFL pro, who'd led them on the field before the game. "Seeing the scores against those small schools, then seeing them doing it against a bigger school in the playoffs, you can't even put words to describe it. There have been great Paradise teams, but nothing like this."

A lifelong Paradise resident was impressed.

"This shows that my town, while it was devastated, we're still here," said contractor Paul Warren, standing against a fence, braced against the cold, cheering on his community. "This is about morale. This is about spirit. It gives us hope."

Dylan Blood looked at all the fans surrounding him—strangers yet friends, former neighbors become family—and he allowed himself the smallest of smiles.

"Last year was so horrible, but things are coming around for all of us. I can feel it in the pulse of the town," he said. "That's who we're playing for: all of them, our families, their families, everybody who still lives in Paradise, everybody who left Paradise. We're showing everyone we've got their backs, and we're always going to have their backs."

Even for those who were not there. *Especially* for those who were not there.

"We are thirty-nine guys scraping and scrapping for thirty thousand who lost their homes, all who died—we are out there running for all of them," said Hartley.

IT WAS SUCH A BIG upset, it literally stopped the Paradise bus. As the Bobcats were beginning their trip home, Prinz received a text on his phone from West Valley coach Greg Grandell, who was following them in his car. Could they pull over at the nearby fire station? He had something to give Prinz.

It turned out that, ordinarily, the top-seeded team was given the first- and second-place sectional plaques in anticipation that the title game would be played at its field. But now that West Valley had lost, custody of those plaques reverted to Paradise. So Prinz stopped the bus, climbed off, and Grandell graciously handed him the plaques and wished him good luck.

"Well, we certainly didn't expect that," said Orlando. "It's kind of awkward. But it's kind of nice."

Sitting in the back of the bus, Brenden Moon was smiling, laughing, filled with the light, the only mark on him a faint impression around his chin. It's where he once wore a necklace containing a tiny urn holding mother's ashes. But the necklace broke, so he'd taken to hanging it on the headboard of his bed, explaining, "It watches over me when I sleep."

Tonight, maybe for the first time since November 8, 2018, Brenden Moon would truly sleep.

The Bobcats were 12-0, and basking in the light.

TWENTY

"Don't let it end . . .
don't let it end."

NOVEMBER 30, 2019

What, though the field be lost, all is not lost;
the unconquerable will and the
courage never to submit or yield.

"Fuck! Fuck! Fuck!"

Players were screaming with passion. Coaches were screaming in anger. This was not how this week was supposed to go. This was not how you prepare for a sectional championship game.

During their final practice before playing heavily favored Sutter Union High School for the prize they'd been building toward all season, the Paradise Bobcats were falling apart. It was freezing. It was dark. The mood was even blacker.

Elijah Gould was literally sitting on teammate Justin Boles as they whaled wild punches at each other. The two tackles cursed. They wrestled. They cursed some more. They finally stopped on their own, as if realizing the absurdity of it all. Boles, a 220-pound junior, stalked off in one direction. Gould walked away in the opposite direction and then circled the field in frustration.

"These guys are finally letting their breath out, all these emotions they're going to have to deal with when the season ends, the reality they have to face—it's all coming out," said Andy Hopper, a worried look crossing his usually jovial face. "But it's too soon. They're exhaling too soon. We still have a championship to win."

Later, while the defense practiced on the field, the rest of the team screwed around on the sideline. There were players lying on the bench, players eating sandwiches, players on their phones, players with no helmets. At one point, several Bobcats played a game where they stood in a circle, then threw up an orange traffic cone and tried to avoid being hit by it. The coaches had been watching the defense on the field until they heard the banging of the cone. They realized they were losing control here. They realized this was not good.

"Will somebody get them a football!" Hopper screamed at the circle of slackers before turning sarcastic. "It's just the sectional championship! It's no big deal! You know you always get in this situation in your life!"

Two other benchwarmers stretched out on the cold grass as if taking a nap, and Nino Pinocchio was enraged.

"Get the fuck up!" he yelled. "You should fucking want to get out there and take somebody's place!"

Now it was Rick Prinz's turn to lose his cool. He knew his team was emotionally exhausted. He understood their need to finally drop this seasonlong façade of adulthood and act like silly kids again. With this being possibly the final practice of the year, he was willing to cut them some slack. But this was too much. This was ridiculous.

"Would you guys get freaking focused!" he screamed. "Get your heads out of your ass!"

The offense took the field and ran the same play three different times. Each time, the defense swallowed up the ball carrier because he had no blockers.

"We've run this play a thousand times, and you still don't know how to down block?!" Prinz hollered in disbelief. "Three times, three different guys. You're not focused, and your mistake is going to cost us tomorrow night."

The players shrugged and looked down. They were tired—too tired to even to be intimidated by their coaches.

"Coach will give us a pep talk tonight, we'll get one before we get on the bus, before we get off the bus, once we get to the game, before the game, halftime, after the game—so many pep talks we can't believe it," Lukas Hartley said with a sigh.

Meanwhile, completing the portrait of chaos, Jeff Trinchera was running around in the freezing temperatures in football pants but no shirt.

"You know what this shows?" said Hopper. "It shows that no matter how much of a burden is placed on their shoulders, they're still just teenagers."

They're still just teenagers. It was an understandable, even comforting notion most days. But one day before those teenagers were being asked to be supermen, that notion scared the hell out of their coaches.

"Maybe this is all finally too much for them," whispered Prinz. "God, I hope it's not too much."

It had been a tough week. It snowed on Tuesday, so they had to practice in the gym. Then on Wednesday they practiced in the snow but were distracted by a catered Wednesday-night Thanksgiving dinner held in the cafeteria. For some, it was their best meal of the week. Prinz couldn't bear to ask them to leave their scattered families on Thanksgiving, so they didn't practice again until today, Friday, the day before the game.

"There's really something off about this week," said Prinz, and he was talking about more than football.

Earlier in the week, Hopper walked outside his home in his flip-flops to retrieve some firewood and slipped in the snow, tearing some ligaments in his knee. Just another tumble in his longest year.

"We always have to keep that fire going, don't we?" he said between limps.

About the same time, across town, Prinz was working on his truck battery when son Seth accidently popped the clutch and knocked him across the driveway.

"*I* may look dumb for what I did, but at least my son *didn't try to kill me,*" Hopper wisecracked.

Through it all, there arose a much bigger warning sign. With rain and a large crowd predicted, officials moved the game from Sutter's potentially dangerous and cramped grass field to the sleek, gray artificial turf of Yuba City's River Valley High. The neutral site, an hour south of Paradise and a mere fifteen minutes east of Sutter, featured spacious stands able to accommodate more Bobcats fans, but the field could be a factor. The Huskies, a fast team, would undoubtedly have an advantage on artificial turf, while the more physical Bobcats preferred mucking it up in the mud. In preparation for the big game, Paradise High's players replicated playing in a rainstorm by marinating footballs in a bucket of water, so they were wet and slippery. Prinz's team wanted slop, but it would have to contend with speed.

"We don't like going to a different field, but there's nothing we can do about it," Prinz said beforehand. "This season has become bigger than all of us."

It had become even larger than Prinz's personal dreams, which he had recently squashed. A couple of weeks earlier, he'd met with school officials at Red Bluff High School about becoming its coach. Red Bluff High was three times the size of Paradise High, with great facilities and probably more money. The word *probably* is used here because Red Bluff never got around to talking salary. Prinz ended the flirtation almost before it started.

"I came to the realization that, at least right now, I can't leave here;

it would be like leaving myself," Prinz said. "It might sound corny, but I don't just live in Paradise. It lives in me."

The importance of his mission was epitomized a week earlier, on the eve of the West Valley game, when he was striding through the campus walkways and feeling stressed to the breaking point. "I was thinking, 'I can't take one more thing; even if a bird poops on my truck, I'm going to lose it,'" he recalled.

Then Prinz turned a corner and, standing there, shaking, was one of his reserve players: a troubled kid who had missed a lot of practice. He approached Prinz and started crying.

"Coach, can we talk?"

Soon the young man was blurting out to Prinz about having lost everything in the fire, about splitting time between living with his mom and his girlfriend, about the anguish of having two households yet, at the same time, having none. Prinz listened, quietly comforted the teenager, and then walked him down the sidewalk before patting him on the back and telling him he'd see him at practice if the kid could make it.

According to Prinz, following the encounter, he looked up and said, "Okay, Lord, I get it. This is where I need to be."

So it was that, a week later, he and his coaches were standing in front of a bunch of shivering, distracted players on the night before the biggest game of their lives, after their worst practice of the season, praying they could find the right words.

Hopper pulled out the two sectional plaques that had been handed to Prinz after the game against West Valley. As you would expect, the championship plaque was significantly larger than the one for the team that finished second.

"You want this one?" Hopper said, holding up the small award before picking up the larger one. "Or do you want *that* one!"

The players roared their selection, but they were treating this more like a parlor game than a pep talk. Paul Orlando stood up and reminded them of the hard stuff. These coaches were privately worried to death about the hard stuff.

"Block and tackle, do the simple things, you do it with all your heart, all your effort. Have attitude, you're a damn good team," he said.

But good enough? Good enough to triumph over not only Sutter, but also a year of misery? Next, Nino Pinocchio took a turn with the motivation.

"Wake up tomorrow morning, take a big ol' pot of water, put it on a stove, put a lid on top of it," he began. "Bubbles start coming to the surface. Then more bubbles. *More* bubbles. By seven, that pot is ready to explode; leave the lid on too long, shit comes off, shit goes everywhere."

That "shit" was supposed to be something good. But what if it was just plain shit? What if the lid had been on since November 8, 2018, and the pot could no longer control what it spewed? Prinz asked himself these questions. In the final pep talk of the night, he tried initially not to inspire them but to calm them.

"I love you guys. I want to keep going on, but tomorrow is just icing on the cake," he said, speaking to them with his hands stuck deep in the pockets of his green parka. "What you've gone through this year, you guys are like men compared to these kids. You don't know it. You don't know how it's made you grow up."

Prinz could have ended the speech there, but he couldn't help himself. He wanted the magic to continue. He knew this town needed the magic to continue.

"The only other thing I have to say is this!" he shouted, and he pulled his hands out of his parka to wave six fingers containing six gleaming sectional championships rings. The players gasped, then broke into a loud cheer.

"You want one of these?" Prinz shouted as his players howled and pointed at the bling. "Look at these! You want one of these?"

The magic was back. Sort of.

"Hey, I want one of those!" Orlando shouted from the back. "Mine are all fucking burned up!"

GAME DAY GAME DAWNED COLD, rainy, and foreboding. Even before the Bobcats boarded the busses for the hourlong ride to Yuba City, they were already chilled and wet. You see, their pregame meal turned out to be a damp picnic. Once again, they found themselves locked out of the cafeteria, this time because anyone who had keys had already left for the game. Because it was a Saturday afternoon, no school officials were around, so they were stuck outside.

"It's been that way all year; our infrastructure has been so affected in so many ways," said Prinz, who didn't have a key.

On a small, covered patio outside the cafeteria, Greg Kiefer, the parent in charge of pregame meals, shrugged and spread out ham, turkey, French rolls, chips, bananas, and pickles on a card table.

"We're just going to have to set it up here," he said. "We've done it before. This is Paradise."

It was forty-three degrees, with a spitting rain. The players tried to eat while sitting on nearby concrete ledges, but the rain was soaking their sandwiches, so they eventually took their meals to the bus.

"This is our program now, we go with the flow," said Andy Hopper. "We're lucky to have food."

The kids had spent all season looking tough. But four hours before kickoff, grabbing their wet lunches while trying to cover their wet heads, they already appeared drenched in defeat.

"Our kids don't shiver," bragged Hopper as the last players boarded the bus.

Oh, but they were shivering when they stepped off the bus at River Valley High. The temperature was dropping, the black sky was opening, and they hustled nervously into a gym, where they sat on a hardwood floor and quietly awaited their fate. They could hear the pounding of the rain outside the doors—pelting, pelting, pelting—as they huddled in tiny circles and stared at one another and searched for one more breath, one more push, one more night.

Some were hunched over, praying. Others were slumped in the corners with their eyes closed. There was no cheering. There were no chants. There was only the sound of that damn rain.

"There's no trophy that can represent the accomplishment of these kids," whispered kicking coach Jeff Marcus, but it was too late for moral victories. The Bobcats needed to leave this place with a win. A season of immense stress and unreal strain had come down to one more shot at redemption for this decimated team and its devastated town.

But the burden that had long fueled them now appeared to be suffocating them.

Prinz made one last appeal to their strength, one last effort at lighting them up, one last pregame pep talk.

"Everything we've been through, it's made you grow up," he told them. "You're men. You're grown men tonight. Those guys didn't lose anything. They didn't lose their house, they didn't lose their town! You had to battle through it, and it made you grow up. You're tougher than them. Let's go prove it! Let's go prove it!"

They cheered, but their voices were low. They lined up to march out of the gym like always, but their cleats were dragging. Like the Sutter High team, they ran to a nearby practice field to warm up early. Yet while the Huskies went back inside later to warm up until the kickoff, Paradise remained outside in the wet chill. Onlookers thought they were crazy. But that was just their way. That was CMF.

"Bottom line, we're just tougher than everybody we play," said Hopper. "That's just who we are."

Sometimes, though, even tough was not enough. Sometimes, perhaps, an entire year of being tough was just too much.

The visiting bleachers were full of rain-soaked Paradise fans showering "CMF!" down upon their wet helmets, a torn town coming together again one more time to cheer for the rebirth.

"I want to be there for the kids; they went through hell," said Michael Bess, a thirty-eight-year-old former Paradise resident. "It's my way of giving back to them or whatever. We just want to show them we're here."

The Bobcats had a chance to reward their wet fans by gaining control early. Sutter gave them some golden opportunities. But they had somehow lost that extra step. They were unable to deliver that extra shove. They were a second slow, a tick off.

In the first half, they recovered a Huskies fumble yet couldn't score. They moved the ball all the way down to Sutter's 6-yard line—but on fourth down, Danny Bettencourt threw an interception. They moved the ball back to the Huskies' 5-yard line in the final seconds of the first half, but Bettencourt took a sack as the clock ran out.

Sutter was more efficient, more controlled. The Huskies scored midway through the second quarter on a 12-yard run by Daniel Cummings and took a 7–0 lead into halftime.

To a Paradise team that should have been leading by two scores, a team that wondered if it had just blown its best chances, the small margin felt huge. As they gathered in one end zone to do their halftime work in the rain, the Bobcats already sounded desperate.

"This is our one shot, right here, right now!" shouted Gould.

"Remember, we're fucking brothers!" shouted Hartley.

Rick Prinz was busy teaching two new plays to giant sophomore Ashton Wagner. That's right, two new plays with possibly just one half left in the season. But right now they needed the six-foot-two, 245-pounder's running strength, even if they had to invent new ways to utilize it.

"We played hard, but we made a lot of mistakes!" shouted Prinz. "Are you cold? Are you cold?"

Their young voices shouted "No!" but their weary eyes and sluggish limbs suggested otherwise. As the second half began, Hank Williams Jr.'s defiant "A Country Boy Can Survive" boomed over the loudspeaker.

But tonight, would he? Could he?

On the first series of the third quarter, Prinz had to call a timeout because some players were out of position. A penalty for a chop block and one for holding slowed down their second drive. The Huskies' Cory McIntyre raced past Dylan Blood for a 32-yard touchdown reception to give Sutter a 14–0 lead.

"Fuck!" shouted Blood, storming to the bench in the pouring rain, crumbling on the bench, bowing his head—a long way from his triumphant moment winning last summer's Glory Hill Run during training camp.

Late in the third quarter, Paradise was finally driving again, and a soaking-wet Hopper was screaming again. His words might have sounded like hyperbole. They were not.

"Run for your life! Run for your life!" he shouted.

Sure enough, the always dependable Tyler Harrison ran for his life and scored from 4 yards out with barely four minutes left in the third quarter. Then lineman Silas Carter blocked a Sutter field goal attempt in the final minutes of the quarter. For all its problems, Paradise still had a chance, down 14–7.

Maybe a country boy really could survive?

Then, trying for the tying touchdown, Bettencourt threw his second interception of the game. The wisdom of this running team suddenly asking its quarterback to do something he'd rarely done all year—the senior had thrown only forty-one passes in the twelve previous games—would be questioned later. For now, the sideline was filled with frustration.

"Don't throw it to them!" begged Prinz. "Don't throw it to them!"

Inspired by the takeaway, the Huskies began moving the ball toward midfield, as Hartley stalked the sidelines, his voice roaring through the foggy mist.

"Don't let it end!" he implored the Bobcats' defensive unit. "Don't let it end!"

Then Sutter High's McIntyre responded with a blazing 54-yard touchdown run through half the defense to make it 20–7. With just nine minutes left in the game, the end was near.

The final blow came with 4:12 on the clock when Hartley was stopped on a fourth-and-1 play to slam the door on the Bobcats' hopes. On a night when the offense combined for 254 rushing yards despite the miserable conditions, it couldn't eke out the one yard it needed.

Hartley left the field staring at the frozen ground. Some players surrounded him in consolation. Other players turned their attention back to the action, screaming for the defense to hold Sutter one more time.

"Give us one more chance! One more chance!" they shouted.

Alas, this was not a fairy tale. This was not some fantasy. There is no such thing as a guaranteed happy ending. This was reality, and sometimes reality can burn you worse than a wall of flames, as the young men of Paradise learned on a night when their season finally ran out of opportunities.

They never got one more chance.

Final score: Sutter Union 20, Paradise 7.

The Bobcats were 12-1, and devastated.

"It was finally all too much," said Andy Hopper, shaking his head as his team helplessly watched the clock tick down to all zeros. "They finally wore down. They finally fell apart."

THE GAME ENDED, AND THE Paradise players and fans spilled out onto the field like always—one team, one town—only this time there was no chanting, only mourning; spectators in ponchos clutching umbrellas and soaked, shivering players huddling together and crying. Tyler Harrison collapsed on one knee. Hartley fell into a teammate's arms.

"It was my fault! It was my fault!" wailed a disconsolate Danny Bettencourt. "Where do I go now? What do we do now?"

Hartley looked up at the rain, falling ever harder, and sobbed loudly.

"I didn't cry this bad when my house burned down," he said.

The wind picked up, it was raining sideways, yet the crowd around the players grew larger, and the noise around them built, and soon there was even cheering and pad pounding, the deep sadness replaced by an eternal gratitude. The town had said it was sorry for the loss, now it wanted to say thank you for the ride.

One woman slipped out the crowd, kissed Rick Prinz on his wet cheek, and disappeared. Another woman, speaking to no one and everyone, said loudly, "For three hours every week, they gave us healing."

The players, however, would not be consoled.

"I hoped it would never end," Hartley said, standing wet in the end zone with his teeth chattering. "I don't want to walk off. I don't want to walk off."

Spencer Kiefer was wandering around on the slick turf as if he were lost. The boy who once found his way down a fiery mountain couldn't navigate off a football field.

"I'm not sure what to do," he said. "I could have played better. I'm not sure what to do now."

Bettencourt, the quarterback with two interceptions, was simply and plainly devastated. The boy who started this journey by having a football in his trunk was now watching it end after he gave that football away, and the grief was overwhelming.

"I cost us the game," he said, his red face streaked with tears.

It was hard to address the players as they hung on to one another and their family members amid the rain, but the coaching staff tried.

"If any of you ever have a dark alley you need to be in the middle of,

I'll follow you every step of the way!" shouted Nino Pinocchio. "You're the toughest sumbitches I've ever met. This is a statement about who you are, who we are, who our town is."

It was also a statement about Pinocchio. It turned out that he had been suffering from a bout of pancreatitis since Thursday. He coached the game while virtually doubled over in pain. He didn't tell anybody. He wasn't going to sit this one out.

"Paradise is who I am," he said simply. "They were not playing this without me."

The hundred or so people clustered around the freezing team were chanting now: "Paradise pride! Paradise pride! Paradise pride."

"No one knows how much they battled," said Prinz. "Just to be in this position was amazing."

As the postgame wake continued, the Bobcats had to fulfill one more duty. They needed to walk to midfield and accept the second-place plaque.

"I'm really, really proud of you guys. Let's go get our award!" shouted Prinz.

Not so fast. Just a day earlier, the players were ridiculing that plaque. They didn't want it. They didn't play for it. They weren't accepting it.

"I'm fucking going home!" shouted Hartley as he spun and started stalking off the field.

His teammates began following him. The Paradise Bobcats had finally had enough. They were finished. They were done. It was over.

No, it was not. You knew it was not.

"Get back here, everybody, *now!*" shouted Prinz. "Own it! Own it! Stay together! Hold it together!"

So they did. They stayed together. They held it together. They owned it. They turned and followed Prinz back to midfield for the losers' ceremony, still hugging, still crying, nobody smiling, but everybody standing steady, surrounding that rejected piece of metal, owning every bit of it.

And thus, on a rainy night soaked in the chill of defeat, the mighty Paradise High School Bobcats completed their rise from the ashes, and city lost became Paradise found.

"It hurts a lot," said Blood. "We wanted to win not only for us but for the whole community—for everyone. But I guess, in a way, we did win. We brought smiles back, we brought families back. For a few hours every week, we made everything seem normal."

As everyone was finally leaving the field, Prinz sighed and said, "I think at some point we thought we were invincible."

Oh, but they were.

TWENTY-ONE

Amen

DECEMBER 19, 2019

*Some natural tears they dropped, but wiped them
soon; The world was all before them, where to choose
their place of rest, and Providence their guide.*

Nineteen days after the end of the football season, it was once again bone-chillingly quiet in what remained of Paradise.

Christmas lights adorned mobile homes squatting on empty lots. Food trucks sat where restaurants once stood. The wind whipped through a darkened high school parking lot. The only sound was sneakers padding slowly on asphalt as the Bobcats trudged into the cafeteria for their season-ending banquet. It's usually held at a catering hall, and is usually a festive affair filled with jokes and hugs. But this year, even that was hard.

They lined up solemnly for the chicken and tri-tip steak and brisket and chili and potatoes au gratin and even alligator. The buffet was paid for by the Penhall Company, a concrete sawing, drilling, and removal firm headquartered in Irving, Texas, through its vice president, Scott Galloway, a member of the 1986 Paradise High team. The teenagers picked their paper plates, utensils, and napkins, then picked the food out of aluminum foil tins.

They separated from their parents and sat at long cafeteria tables as if it were just another lunch period. They bowed their heads, they swallowed huge bites of the food, and, together, they grieved.

"I don't want to touch my pads. I can't touch my pads," said Dylan Blood. "I honestly think we wore out. Going to practice every day even though we didn't have a house, didn't even have cleats, we were emotionally drained. It still hurts. A lot."

Spencer Kiefer couldn't even bring his soiled jersey and pants into his house. The uniform was still lying in a dirty heap on his back patio.

"I'm not going near it," he said. "I don't want anything to do with it."

Danny Bettencourt was still angry, not only at himself but also at the world. Having had three weeks to process the loss, he now regretted accepting all of the blame.

"I wish I could take it back, saying that it was my fault," he said. "It was everybody's fault. It was just fate. We were all shaking out there; I don't know if it was from the cold or the nerves. It was just a tough way to end."

Stetson Morgan, though, was in no mood for excuses.

"Could have, should have, would have—we just ran out of gas," he said.

Kasten Ortiz slept the entire day after the Saturday-night loss and still couldn't deal with his pain.

"It was a helluva ride, every Friday night, people from Paradise in the stands. We were that beacon of hope for them, and now . . . I'm in disbelief that I will never set foot on that field again," said the high school senior. "It kills me that I never get to practice again."

He said he still thought about muscling an opposing defensive lineman six yards downfield to open a hole for the running backs. He still thought about making a pancake block and (as its name would suggest) flattening a linebacker trying to get at the quarterback. He still remembered all the plays—the 28 sweep, the tackle trap, the 36 down—and reenacted his resilience in his head. He constantly replayed in his mind the games from the season that made him a man.

"I can still feel a running back coming off my butt and sprinting to the end zone," Ortiz said. "Clearing a path for someone else—nothing like it."

Elijah Gould was preparing for knee surgery and dreading the new reality of life without football. It was a reality of temporary housing and an uncertain future, one that they'd spent the last several months trying to avoid.

"Football distracted everyone for so long, it was so great," he said. "It's hitting us more now, it's all hitting us at once, and that's a lot."

Greg Kiefer watched his son and teammates from another table and shook his head. He knew these kids would be hurt by the loss to Sutter High, but he never imagined they would still be hurting so bad all this time later.

"They're all still so sad," he said. "I tell them they brought this town together, but they don't care, they just wanted to win. They won't realize until later on what they did."

The remorse engulfed more than just the players. Coach Rick Prinz had not slept in days, tossing and turning and wondering how he could have guided them better. He thought about the terrible practice the day before the championship showdown. He wondered if he blew it.

"The Friday practice was a joke," he said bluntly. "They kept making assignment errors. It should be a day of no mistakes; we had kids not focused. I was trying to be patient, but it was building up inside of all of us. The team was disintegrating, and it was too late. I didn't know how to get it back on course."

He thought about one running play that epitomized the entire Sutter game. A veteran offensive lineman pulled instead of blocking head-on, resulting in ball carrier Lukas Hartley getting crushed at the line of scrimmage. It was that fourth-down play in the fourth quarter, with four minutes and change left on the clock, when they needed a yard—a mere three feet—and fell short. Game over.

"It was a basic play, the kid has run the play for two years, and he just forgot," said Prinz. "I saw it at that last practice. I'm going to change things next year. Our last practices before games are going to be a lot tougher."

Just like the players, the coaches also wore down. In talking to their team two days after the loss in a postgame, postseason autopsy in classroom 114, they were raw. They tried to convince the players that this football season wasn't really about football.

"I don't think I fully understand what we've done for our town. I don't understand the enormity of the story, but you need to appreciate it," Prinz told them. "It really wasn't about a championship, it was about, week after week, giving our community something to rally around."

Paul Orlando admitted that he was among those who needed the sport for that very reason.

"I just want to thank you for making my life easier," the defensive coordinator and school handyman told them. "Coming to football practice with you guys every day, it made my life ten times easier, and now that football is over, I'm out there putting up soccer goal posts and putting on soccer nets, and I don't even know what a soccer ball is. I'm having a difficult time dealing with that. This really had nothing to do with winning games, it was about focusing on football and not thinking about living in a twenty-five-foot trailer."

Nino Pinocchio agreed that this season was about giving everyone, especially the homeless coaches, a reason to believe in hope.

"Like you, I went through a lot of shit, and this has given me a reason every day to get up and do something," he said. "I'm not good at a lot of things in life, but I'm okay with this, coaching; it's one thing I can do to help the healing. I loved coming up to my hometown every day, and I won't have a reason to do that tomorrow, and that really sucks."

What doesn't suck, said Pinocchio, is the tale they told.

"You lost everything you owned, and most communities wouldn't have been able to put together a football team, much less a twelve-and-one team," he said. "You built a story. You built a story that people will be talking about for years. Your kids will not only ask you, 'You lived in Paradise when it burned down, what was it like?' They'll also say, 'Dad, you were part of that football team, what was that like?' Nobody will be able to tell the story like you. You lived it. You've done something no one else has ever done. You should be so proud."

And with that, Pinocchio quietly returned to coaching retirement, leaving his colleagues in awe at his selfless contribution.

"I don't know that we would have rebuilt this program without Nino," said Prinz appreciatively. "In many ways, he was our heart and soul."

Pinocchio was known for that trademark camouflage bucket hat. There are bits of blood, chunks of dirt, and aging sweat stains on that hat. Pinocchio wore it nearly every day during the 2019 season to remind him that this resurrection required all three.

"I was maybe a small part of proving to the world that the toughest kids in the world come from Paradise football," Pinocchio said. "It was my honor."

Also eventually retired was Shannon Magpusao's roving shelter. Six months after the 2019 season, that rattling white Dodge minivan with more than two hundred thousand miles on it finally died. It was left in the driveway of Magpusao's brother's house as a monument to caring.

"I can't get rid of it," said Magpusao, cornerstone, inspiration, Samaritan, saint.

Slowly, in little moments around town, even through their grief, the players started to realize their impact.

A COUPLE OF DAYS AFTER the loss, senior Dylan Blood attended a birthday party, and suddenly all of his mom's friends were coming up to him, shaking his hand, thanking him.

"Why are you thanking me? We didn't win it."

"It didn't matter," one told him. "You played. You gave us a chance to come to the games. We saw our friends who we've been missing since the fire. You gave us a piece of our lives back."

Late in the season, Josh Alvies experienced the same gratitude when he went to one of the few reopened Paradise markets, and the woman behind the deli counter recognized him even though he was a nondescript lineman.

"Hey, you're a football player, aren't you?" she said. "We're live-streaming the games. Thank you for helping bring our town back."

Later, on December 11, the Paradise Town Council later honored the team with a proclamation at Paradise Town Hall.

"They've been a rallying point for a lot of people," said Greg Bolin, Paradise's former mayor. "They didn't give up, they didn't stop, they fought through adversity. They've been an example of what our whole town is doing now."

It turned out to be their only official celebration, unfortunately. Three months after the Town Hall proclamation, Paradise, along with most of the world, was in a sort of suspended animation due to the COVID-19 pandemic that was raging through community after community. A planned spring parade to honor the Bobcats had to be canceled, as was a California Coaches Association banquet at which Rick Prinz was to be officially named the organization's High School Football Coach of the Year. A year after the Camp Fire, the pandemic robbed the Paradise students of even their remaining shreds of normalcy, no prom, drive-through graduation, unfairly piling tough upon tough.

And so, the December 19 banquet was pretty much the Paradise Bobcats football team's only moment of collective reflection and gratitude.

"You have a lot of people rooting for you out of state. I travel a lot, and people everywhere say, 'Paradise? Oh my God, those football players!'" said the first speaker, alumnus Nikki Roethler, class of 1987, one of about a dozen alumni in attendance.

There was a highlight video, during which somebody farted loudly, and everybody laughed. Well, not everybody. Hopper was crying. After the video ended, he stood up and said simply, "I want you guys to know, all of you saved my life."

The first honoree was *Los Angeles Times* video journalist Rob Gourley, who had been documenting the season. In typical Paradise fashion, he was not being honored for his art but for his brotherhood.

"Stand up, Rob, turn around, pull up your shirt! Guess what we did to the city boy!" shouted Hopper.

Gourley raised his shirt in front to reveal a small tattoo on his chest. You guessed it: "CMF."

Gourley grinned. The players gasped, then howled, then stomped their feet in the loudest accolade of the night.

Various coaches' awards were given to Brenden Moon, Danny Bettencourt, and junior lineman Christian Zausch. Of Moon, Pinocchio said, "I will forever have a place in my heart for this young man, because in my mind, what he did against West Valley was a perfect example of what 'CMF' and Paradise football are about."

Ben Weldon won an award, but he wasn't there. Neither was Tyler Harrison, another award recipient. School ended early on this particular Thursday, and because some players' new homes were so scattered and distant, it was tough for everyone to drive back.

"Just another thing, always one more thing," said Prinz.

Luke Hartley, huddled under a John Deere hoodie, walked up to receive his honor. He had finally grown quiet. Kasten Ortiz was honored for his smarts. Blake White was honored for his stubbornness. Remember when he refused to leave the car on the night the San Francisco 49ers honored both Paradise and Pleasant Valley High? These Bobcats would always remember.

"He sent a message that we don't want anybody to feel sorry for us," said Hopper. "That was huge."

Outside linebacker Tyler Hooks, a senior, was honored. So was prayerful Alvies and fearless Kiefer and giant Ashton Wagner. Some of them were wearing hoodies that read "Embrace the Suffering." Prinz worried that for many, the suffering was just beginning.

Away from the microphone, he said quietly, "My big concern is that . . . now that the season is over . . . they've lost their family. The tears I saw after the game, I guarantee you, they knew they lost their family. Many of them have said, 'Now what are we going to do?' And they mean it. Now we have to face reality. Now we have to face life postfire. They're no longer in that bubble. None of us know how to go about it, honestly. We could all use some help."

For now, they would eat their tri-tip and laugh at farts and cling to the last vestiges of their brotherhood, while Paul Orlando would honor them with a rare show of emotion.

"These kids won everything for us first day they showed up for practice," he said, his gravelly voice growing thick. "They made me want to do life again."

Hopper honored them all with a promise of hope.

"Every one of these boys deserves so much more for what they've done for this town," he said. "You've done something great for community, and it's going to continue on. What we learned this year is bigger than any of us and will last forever."

What also will last forever are hard-earned high school diplomas. Amazingly, all twenty seniors on the 2019 Paradise Bobcats graduated later that school year. This included Jeff Trinchera, who guaranteed that graduation by joyously nailing a spring technology test that jumped his final grade from a D-minus to a C. Yet he couldn't attend graduation.

He was needed at his construction job, working on houses and continuing his mission.

"I'm rebuilding Paradise," he said.

What the future holds for the Paradise football program is anybody's guess. Two years after the fire, fewer than five hundred new homes had been built in a town that lost around fourteen thousand. Enrollment had dropped from a postfire figure of around 590 to approximately 430 in 2020, as parents grew weary of making the long drive up the mountain. The varsity football team numbered just 34 in 2020, 5 fewer than even in the postfire 2019. Because of the pandemic they didn't start the 2020 season until March of 2021, and finished an abbreviated schedule with a 3-2 record. Fewer students, and players, are expected in the coming years.

"The euphoria of coming back to the high school is dying," said coach Jeff Marcus, who had to cut short his second go-round as principal in the fall of 2019 due to a California State Teachers' Retirement System rule. "People driving ninety minutes to get here, that won't endure. Students walking thirty minutes to a bus stop for an hourlong ride, that's not going to keep happening. And the lack of general contractors, the lack of funds, the red tape, it's all slowing the process of people moving back here. Since the fire, everything is still working against us."

Anne Stearns appeared at the banquet to impress upon the kids that, despite some of the numbers, they really had made a positive impact and would continue to do so. "We're trying to make them understand, this is not where the road ends, this is the beginning," said the athletic director. "I'm not sure of anything, but I do know I want a new weight room, team room, turf, and a new scoreboard."

Thanks to the football team, she may actually get some of that. While the Bobcats didn't ultimately triumph on the field, their inspirational story filled the coffers of the booster club. Shawn Meaika, founder of the Family First Life marketing organization, donated $75,000, and two business associates chipped in $25,000 each. Then there were the equally meaningful gestures from fans who would walk up to Coach

Prinz after games and slip money in his pockets, sometimes as much as $500. A couple of high schools—Casa Roble Fundamentalist, in Orangevale, and Red Bluff—donated to the players gift cards ranging from $100 to $200. There were also monetary blessings from an anonymous group that asked to meet Prinz at a Starbucks near his temporary home in Willows. Once there, they gave him $7,000 in cash to distribute to the neediest players, which he did: seven players, $1,000 apiece. The booster bank account usually held $10,000. By the end of the 2019 season, it had a balance of $140,000. The school quickly used some of it to buy new weights, with more improvements coming.

"I don't know, maybe they can build us a little room so we don't have to hold halftime in a dark field?" said Prinz.

In front of a drowsy, drained banquet audience, it was time for the coach's last speech. He had cried enough. He had coached enough. The old youth pastor decided to deliver something that sounded like a prayer. He pulled a wrinkled scrap of paper out of his pocket and began.

"On 11/8/18 at 8:34 a.m., I sent this text to our football team," he said. The plan is to practice at 3:00 today. If it is to Smokey, we will modify our activity. I will keep you informed if anything changes.

He paused.

"Eleven minutes later, we were running for our lives," he said. "These young men faced the reality of death, they faced real stresses in life that a lot of young people don't have to face, they lost everything, they lost their possessions, their home, and their town, they're living with relatives, in hotels, trailers, and cars. Some were even homeless."

The cafeteria grew quiet. Heads were bowed. Hands were clasped. The story of the 2019 Paradise Bobcats was being told in this invocation. Perhaps Prinz wasn't done crying after all. He cleared his throat and continued.

"I could see the anguish and fear in my players' eyes," he said. "We didn't have a school, we didn't have a practice field, we didn't have cleats, *we didn't even have a football!*"

He was shouting now, shouting down the demons, shouting out the pain, honoring the strength that led to the gift that guided them out of the flames and into the greatest season of their lives.

"But we had each other!"

Amen.

ACKNOWLEDGMENTS

Just like the 2019 Paradise Bobcats football team, this book had a coach. It was the same coach. Rick Prinz taught me, inspired me, and guided me throughout every step of this journey. He answered every call, every text, every question, always. He unselfishly shared of himself while living with ramifications from the Camp Fire and enduring the pandemic shutdown. He held me together while trying desperately to hold his football team together. He keeps telling the wonderful Veronica that he is going to retire, and I know he deserves to retire, but the thought of him no longer walking the Paradise sidelines is unimaginable. I hope he never stops coaching. The universe needs to continue to be touched by good souls like him. Not to mention, Paradise needs more late-Friday-night parties on his back porch. Thanks, Old Man Prinzer. This book would have been utterly impossible without you.

Thanks also to Prinz's tremendous Paradise coaching staff from 2019, they steered me with good humor and patient grace . . . Andy Hopper, Nino Pinocchio, Paul Orlando, Bryson Baker, Shannon Magpusao, Bobby Richards, John Wiggins, Jeff Marcus, Weston Griffiths, and trainer Chip Schuenemeyer. When I needed the big picture, I relied on the indomitable Paradise High athletic director Anne Stearns, who was smart and kind and tough while incredibly handling arguably the biggest administrative challenge in high school sports history.

Then there were the Paradise players from 2019, their names are listed throughout this book, I owe them my deepest gratitude for enduring my nagging presence and entertaining my dumb questions with wisdom far beyond their years. They showed me that powerful faith and endless hope looks like a determined teenager who keeps hollering and hitting even when everything is telling him to quit. It looks like someone who, after losing everything, still believes he is going to win. Thank you to all the Bobcats for proving that even a fire of historically damaging proportions cannot burn a passion for life.

A special thanks to the citizens of Paradise, especially the players' parents and relatives, many of whom I would see at the football games cheering for not just the team, but the town. I've never met a more resilient population, their city will come back, Paradise Strong, bet on it. To those good folks at Starbucks who always remembered my order, save the name. A shout-out also to Sharon Martin and Rick Silva, two strong presences roaming the Paradise sidelines, thanks for the spots, thanks for the stats, thanks for the words.

The genesis for this project began when the great Gary Pine of Azusa Pacific introduced me to the great Rick Prinz. It was Pine's best story idea in a lifetime of cool story ideas. The idea was then expanded when the *Los Angeles Times*'s Angel Rodriguez allowed me the crazy freedom to write, over a span of seven months, a series of nine stories on a rural high school football team located 470 miles north of Los Angeles. Thank you forever, Angel. Those stories were edited by the tremendous *Times* copy desk after being massaged by Mike Hiserman. You the man, Mike. The thought that these stories could actually become a book was planted in my head by colleagues Nathan Fenno and David Wharton, two brilliant minds connecting with my simple one. My *Times* family is surely one of the most tightly knit sports staffs in the country, thanks for always having my back. The vision for the book was then honed by the great work of *Times* photographer Wally Skalij, you should have won a Pulitzer, bro. The final approval for the book came from the *Times*'s Scott Kraft, who was gracious and understanding and took all of two seconds to say yes.

The book then became a reality through the great work of eternally encouraging agent Susan Canavan of the Waxman Literary Agency.

Susan, even when I'm struggling to find the words, you always make me feel I can write anything. The book was shepherded by great Harper-Collins editors Mauro DiPreta and Nick Amphlett—Nick, your heavy lifting was so impressive—and amazing copy editor and writer Philip Bashe. As I started the work, I was inspired by my dear friend Gene Wojciechowski, The World's Smartest Man, even though I have the trophy. And the actual time to write the book? Thank you to the *Times*'s Chris Stone for understanding beyond all understanding.

During this project, I was stricken with COVID-19. It caused huge headaches, both literally and figuratively, and required great patience from everyone involved. I will never forget the understanding shown to me by Susan and the folks at HarperCollins, and will be forever appreciative of their kindness, concern, and their willingness to give me one long leash.

Because of my illness, this book required even more inspiration from my family, who I could thank in 72,000 more words and it wouldn't be enough. Start with my mother, Mary Plaschke, my daily touchstone, my eternal rock, my constant encouragement, the crazy believer in all things me. Guess what, Mom? I can finally answer your yearlong daily questions about whether I had finished this extraordinary journey. Yes! We did it! My three children are quite simply my life, and all three constantly cheered me on, thank you Tessa, Willie, and MC. The laughs came from Tessa, the John Milton quotes came from Willie, and the history lesson came from MC. I want to also thank my brothers Brad and Bob and sister Beth for always saying, "How's the book going?" with the greatest of hope, while brother Andrew Ladores asked the identical question while managing our Billfish teams and Sam Umfress did the same over margaritas.

Finally, the strength and soul of this project belongs to my love, my partner, my best friend, the incomparable Roxana Verano. She was the first person I told when I started the book. She was the first person I told when I finished the book. That is not coincidence. She is the vision behind my thoughts, the voice behind my words, the power behind my work, and the one who left food outside my front door every day during my bout with COVID. You see, honey? I knew you were there. You're always there. You are my paradise found.

Bill Plaschke
June 1, 2021